Major Bill Dampsell
Woodfield Farm
Cazenovia

December 1960

A ZOO IN MY LUGGAGE

GERALD DURRELL

A Zoo in My Luggage

With illustrations by
RALPH THOMPSON

THE VIKING PRESS
NEW YORK
1960

Published in 1960 by The Viking Press, Inc.

625 Madison Avenue, New York 22, N. Y.

Library of Congress catalog card number: 60-14087

Printed in the U.S.A. by The Colonial Press Inc.

CONTENTS

A ZOO IN MY LUGGAGE

A WORD IN ADVANCE

This is mainly the chronicle of a six-month trip that my wife and I made to Bafut, a mountain grassland kingdom in the British Cameroons in West Africa. Our reason for going there was, to say the least, a trifle unusual. We wanted, quite simply, to collect our own zoo.

Since the end of the war I had been financing and organizing expeditions to various parts of the world to collect wild animals for various zoological gardens. Bitter experience over the years had taught me that the worst and most heartbreaking part of any collecting trip was at the end when you had to part with the animals upon which you had lavished care and attention for months. After being mother, father, food provider, and danger eradicator to an animal for six months, you found that you had built up a very real friendship with it. The creature trusted

3

you and, what is more important, behaved naturally when you were around. Then, just when this relationship should have started to bear fruit, when you should be in an unique position to study the animals' habits and behaviour, you were forced to part with it. There was only one answer to this problem, as far as I was concerned, and that was to have a zoo of my own. Then I could bring my animals back knowing what type of cages they were going to inhabit, what sort of food and treatment they were going to receive (a thing which you could not, unfortunately, be sure about with some other zoos) , and secure in the knowledge that you could go on studying them to your heart's content. The zoo, of course, would have to be open to the public, so that, from my point of view, it would be a sort of self-supporting laboratory in which I could keep and watch my animals.

There was another and, to my mind, more urgent reason for creating a zoo. I, like many other people, have been worried by the fact that year by year, all over the world, various species of animal life are being slowly extermin-ated in their wild state, directly or indirectly by the inter-ference of mankind. While there are many worthy and hard-working societies all over the world tackling this problem, there are, I know, a great number of animal species which, owing to the fact that they are small and generally of no commercial or touristic value, are not re-ceiving adequate protection. To me the extirpation of an animal species is a criminal offence, just as the destruction of something else that we cannot recreate or replace, such as a Rembrandt or the Acropolis, would be. In my opinion zoological gardens all over the world should have as one of their main objectives the establishment of breeding colonies of such rare and threatened species. Then, should what seems to be the inevitable happen and the animal become extinct in the wild state, at least it is not lost completely. I had, for many years, wanted to start a zoo with just such an object in view, and now it seemed the ideal moment to begin.

Of course, any normal person smitten with such an ambition would have got the zoo first and the animals next. But throughout my life I have rarely if ever achieved what I wanted by tackling it in a logical fashion. So, naturally, I went and got the animals first and then set about the task of finding my zoo. This was not so easy as it might seem on the face of it, and looking back on it now I am speechless at my audacity in trying to achieve success in this way.

So this is the story of my search for my zoo, and it explains why, for some considerable time, I had a zoo in my luggage.

MAIL BY HAND

From my seat on the bougainvillaea-enshrouded veranda I could look out over the blue and glittering waters of the bay of Victoria, a bay dotted with innumerable tiny, forest-encrusted islands like little green, furry hats dropped carelessly on the surface. Two grey parrots flew swiftly across the sky, wolf-whistling to each other and calling "Coo-eee" loudly and seductively in the brilliant blue sky. A flock of tiny canoes, like a school of black fish, moved to and fro among the islands, and dimly the cries and chatter of the fishermen came drifting across the water to me. Above, in the great palms that shaded the house, a colony of weaver birds chattered incessantly as they busily stripped the palm fronds off to weave their basket-like nests, and behind the house, where the forest began, a tinker bird was giving its monotonous cry, "Toink, toink, toink," like someone beating on a tiny anvil. The sweat was running down my back, staining my shirt black, and the glass of beer by my side was rapidly getting warm. I was back in West Africa.

Dragging my attention away from a large orange-headed lizard that had climbed onto the veranda rail and was busily nodding its head as if in approval of the sunshine, I turned back to my task of composing a letter.

The Fon of Bafut,
Fon's Palace,
Bafut,
Bemenda Division, British Cameroons

I paused here for inspiration. I lit a cigarette and contemplated the sweat marks that my fingers had left on the keys of the typewriter. I took a sip of beer and scowled at my letter. It was a difficult screed to compose for a number of reasons.

The Fon of Bafut was a rich, clever, and charming potentate who ruled over a large grassland kingdom in the mountain area of the north. Eight years previously I had spent a number of months in his country in order to collect the strange and rare creatures that inhabited it. The Fon had turned out to be a delightful host, and we had many fantastic parties together, for he was a great believer in enjoying life. I had marvelled at his alcoholic intake, at his immense energy, and at his humour, and when I returned to England I had attempted to draw a picture of him in a book I wrote about the expedition. I had endeavoured to show him as a shrewd and kindly man, with a great love of music, dancing, drink, and other things that make life pleasant, and with an almost childlike ability to enjoy himself. Now I wanted to revisit him in his remote and beautiful kingdom and renew our friendship, but I was a little bit worried. I had realized—too late—that the portrait I had drawn of him in my book was perhaps open to misconstruction. The Fon might well have thought that

the picture I had drawn was that of a senile alcoholic who
spent his whole time getting drunk amid a bevy of wives.
So it was with some trepidation that I sat down to write to
him and find out if I would be welcome in his kingdom.
That, I reflected, was the worst of writing books. I sighed,
stubbed out my cigarette, and started.

My dear friend,
    *As you may have heard I have returned to the Cameroons
in order to catch more animals to take back to my country.
As you will remember when I was last here I came up to
your country and caught most of my best animals there.
Also we had a very good time together.*
    *Now I have returned with my wife and I would like her
to meet you and see your beautiful country. May we come
up to Bafut and stay with you while we catch our animals?
I would like to stay once more in your rest house, as I did
last time, if you will let me. Perhaps you would let me
know?*

Yours sincerely,
Gerald Durrell

    I sent this missive off by messenger together with two
bottles of whisky which he was given strict instructions
not to drink on the way. Then all we could do was to wait
hopefully, while our mountain of luggage smouldered under tarpaulins in the sun, and the orange-headed lizards
lay dozing on top of it. Then after a week the messenger
returned and drew a letter out of the pocket of his tattered
khaki shorts. I ripped open the envelope hastily and spread
the letter on the table, where Jacquie and I craned over
it.

Fon's Palace,
Bafut, Bemenda.
25th January 1957.

My good friend,
    Yours dated 23rd received with great pleasure. I was more than pleased when I read the letter sent to me by you, in the Cameroons again.
    I will be looking for you at any time you come here. How long you think to remain with me here, no objection. My rest house is ever ready for you at any time you arrive here.
    Please pass my sincere greetings to your wife and tell her that I shall have a good chat with her when she comes here.

Yours truly,
Fon of Bafut

PART ONE: EN ROUTE

MAIL BY HAND

To the Zoological Officer,
U.A.C. Managers House,
Mamfe

Dearest Sir,
I have once been your customer during your first tour of the Cameroons and get you different animals.
I send here one animal with my servant, I do not know the name of it. Please could you offer what price you think fit and send to me. The animal has been living in my house almost about three weeks and a half.
With love sir.

I am,
Yrs. sincerely,
Thomas Tambic. Hunter

1 : THE RELUCTANT PYTHON

I had decided that, on the way up-country to Bafut, we would make a ten-day stop at a town called Mamfe. This was at the highest navigable point of the Cross River, on the edge of an enormous tract of uninhabited country, and on the two previous occasions when I had been to the Cameroons I had found it a good collecting centre. So we set off from Victoria in an impressive convoy of three trucks, Jacquie and myself in the first, our young assistant Bob in the second, and Sophie, my long-suffering secretary, in the third. The trip was hot and dusty, and we arrived at Mamfe in the brief green twilight of the third day, hungry, thirsty, and covered from head to foot with a fine film of red dust. We had been told to contact the United Africa Company's manager on arrival, and so our trucks roared up the drive and screeched to a halt outside a very impressive house, ablaze with lights.

The house stood in what was certainly the best position in Mamfe. It was perched on top of a conical hill,

one side of which formed part of the gorge through which
the Cross River ran. From the edge of the garden, fringed
with a hedge of the inevitable hibiscus bushes, you could
look straight down four hundred feet into the gorge, to
where a tangle of low growth and taller trees perched precariously on thirty-foot cliffs of pleated granite, thickly
overgrown with wild begonias, moss, and ferns. At the foot
of these cliffs, round gleaming white sandbanks and
strange, ribbed slabs of rock, the river wound its way like
a brown, sinuous muscle. On the opposite bank there were
small patches of farmland along the edge of the river, and
after that the forest reared up in a multitude of colours
and textures, spreading endlessly back until it was turned
into a dim, quivering, frothy green sea by distance and
heat haze.

I was, however, in no mood to admire views as I uncoiled myself from the red-hot interior of the truck and
jumped to the ground. What I wanted most in the world
at that moment was a drink, a bath, and a meal, in that
order. Most important of all, I wanted a wooden box to
house the first animal we had acquired. This was an extremely rare creature, a baby black-footed mongoose,
which I had purchased from a native in a village some
twenty-five miles away when we had stopped there to buy
some fruit. I had been delighted that we had started the
collection with such a rarity, but after struggling with her
for two hours in the front seat of the truck, my enthusiasm
had begun to wane. Firstly she had wanted to investigate
every nook and cranny in the cab, and, fearing that she
might get tangled up in the gears and break a leg or something, I had imprisoned her inside my shirt. For the first
half-hour she had stalked round and round my body,
sniffing loudly. For the next half-hour she had made several

determined attempts to dig a hole in my stomach with her exceedingly sharp claws, and on being persuaded to desist from this occupation she had seized a large portion of my abdomen in her mouth and sucked it vigorously and hopefully, while irrigating me with an apparently unending stream of warm and pungent urine. This did not add anything to my already dusty and sweaty appearance. So, as I marched up the steps of the U.A.C. manager's house, with a mongoose tail dangling out of my tightly buttoned, urine-stained shirt, I looked, to say the least, slightly eccentric. Taking a deep breath and trying to look nonchalant, I walked into the brilliantly lit living room, and found three people seated round a card table. They looked at me with a faint air of inquiry.

"Good evening," I said, feeling rather at a loss. "My name's Durrell."

It was not, I reflected, the most telling remark made in Africa since Stanley and Livingstone met. However, a small, dark man rose from the table and came towards me, smiling charmingly, his long black hair flopping down over his forehead. He held out his hand and clasped mine, and then, ignoring my sudden appearance and my unconventional condition, he peered earnestly into my face.

"Good evening," he said. "Do you by any chance play canasta?"

"No," I said, rather taken aback. "I'm afraid I don't."

He sighed, as if his worst fears had been realized. "A pity . . . a great pity," he said; then he cocked his head on one side and peered at me closely.

"*What* did you say your name was?" he asked.

"Durrell . . . Gerald Durrell."

"Good heavens," he exclaimed, "are you that animal maniac head office warned me about?"

"I expect so."

"But my dear chap, I expected you two days ago. Where have you been?"

"We would have been here two days ago if our truck hadn't broken down with such monotonous regularity."

"These local trucks are bloody unreliable," he said, as if letting me into a secret. "Have a drink?"

"I should love one," I said fervently. "May I bring the others in? They're all waiting in the trucks."

"Yes, yes, bring 'em all in. Of course. Drinks all round."

"Thanks a lot," I said, and turned towards the door.

My host seized me by the arm and drew me back. "Tell me, dear boy," he said in a hoarse whisper. "I don't want to be personal, but is it the gin I've drunk or does your stomach *always* wriggle like that?"

"No," I said gravely, "it's not my stomach. I've got a mongoose in my shirt."

He gazed at me unblinkingly for a moment. "Very reasonable explanation," he said at last.

"Yes," I said, "and true."

He sighed. "Well, as long as it's not the gin I don't mind *what* you keep in your shirt," he said lavishly. "Bring the others in and we'll kill a noggin or two before you eat."

So we invaded John Henderson's house, and within a couple of days we had turned him into what must have been assuredly the most long-suffering host on the West Coast of Africa. For a man who likes his privacy to invite four strangers to live in his house is a noble deed to start with. But when he has no liking for, and a grave mistrust of, any form of animal life, to invite four animal collectors to stay is an action so heroic that no word can describe it. So, within twenty-four hours of our arrival there was not only a mongoose, but a squirrel, a bushbaby, and two monkeys living on the veranda of John's house.

While John was getting used to the idea of having his legs embraced by a half-grown baboon every time he set foot outside his own front door, I sent messages to all my old contacts among the local hunters, gathered them together, and told them the sort of creatures we were after. Then we sat back and awaited results, but they were some time in coming. Then, early one afternoon, a local hunter called Agustine appeared, padding down the drive, wearing a scarlet and blue sarong and looking, as always, like a neat, eager Mongolian shopwalker. He was accompanied by one of the largest West Africans I have ever seen, a great scowling man who must have been at least six feet in height, and whose skin—in contrast to Agustine's golden bronze shade—was a deep soot black. He clumped along beside Agustine on such enormous feet that at first I thought he was suffering from elephantiasis. They stopped at the veranda steps, and while Agustine beamed cheerily his companion glared at us in a preoccupied manner, as though endeavouring to assess our net weight for culinary purposes.

"Good morning, sah," said Agustine, giving a twist to his highly coloured sarong to anchor it more firmly round his slim hips.

"Good morning, sah," intoned the giant, his voice sounding like the distant rumble of thunder.

"Good morning. You bring beef?" I inquired hopefully, though they did not appear to be carrying any animals.

"No, sah," said Agustine sorrowfully, "we no get beef, I come for ask Masa if Masa go borrow us some rope."

"Rope? What do you want rope for?"

"We done find some big boa, sah, for bush. But we no fit catch um if we no get rope, sah."

Bob, whose speciality was reptiles, sat up with a jerk. "Boa?" he said excitedly. "What does he mean . . . boa?"

"They mean a python," I explained. One of the most confusing things about pidgin-English from the naturalist's point of view was the number of wrong names that were used for various animals. Pythons were boas, leopards were tigers, and so on. Bob's eyes gleamed with a fanatical light. Ever since we had got on board the ship at Southampton his conversation had been almost entirely confined to pythons, and I knew that he would not be really happy until he had added one of these reptiles to the collection.

"Where is it?" he asked with ill-concealed eagerness.

" 'E dere dere for bush," said Agustine, waving a vague arm that embraced approximately five hundred square miles of forest. " 'E dere dere for some hole inside ground."

"Na big one?" I asked.

"Wah! Big?" exclaimed Agustine. " 'E big too much."

" 'E big like dis," said the giant, slapping his thigh, which was approximately the size of a side of beef.

"We walker for bush since morning time, sah," explained Agustine. "Den we see dis boa. We run quick-quick, but we no catch lucky. Dat snake get power too much. 'E done run for some hole for ground and we no get rope so we no fit catch um."

"You done leave some man for watch dis hole," I asked, "so dis boa no go run for bush?"

"Yes sah, we done lef' two men for dere."

I turned to Bob. "Well, here's your chance, a genuine wild python holed up in a cave," I said. "Shall we go and have a shot at it?"

"God, yes! Let's go and get it right away," exclaimed Bob, in the fervent tones of a Crusader who has been offered the opportunity of obtaining a fragment of the true cross.

I turned to Agustine. "We go come look dis snake, Agustine, eh?"

"Yes, sah."

"You go wait small time and we go come. First we get rope and catch net."

So, while Bob hurried out to our pile of equipment to fetch rope and nets, I filled a couple of bottles with water and rounded up Ben, our animal boy, who was squatting outside the back door, flirting with a damsel of voluptuous charms.

"Ben, leave that unfortunate young woman alone and get ready. We're going for bush to catch a boa."

"Yes, sah," said Ben, reluctantly leaving his girl friend. "Which side dis boa, sah?"

"Agustine say it's in a hole for ground. That's why I want you. If this hole is so small that Mr. Golding and I no fit pass you will have to go for inside and catch the boa."

"Me, sah?" said Ben.

"Yes, you. All alone."

"All right," he said, grinning philosophically, "I no de fear, sah."

"You lie," I said. "You know you de fear too much."

"I no de fear, for true sah," said Ben in a dignified manner. "I never tell Masa how I done kill bush cow?"

"Yes, you told me twice, and I still don't believe you. Now, go to Mr. Golding and get the ropes and catch nets. Hurry."

To get to the area of country in which our quarry was waiting, we had to make our way down the hill to the river bank, and then cross the river by means of the ferry. This consisted of a large, banana-shaped canoe which appeared to have been constructed about three centuries ago, and to have been deteriorating slowly ever since. It was paddled by a very old man who looked in immediate danger of dying of a heart attack, and he was accompanied by a small boy whose job it was to bail out. This was somewhat of an unequal struggle, for the boy had a small rusty tin for the job, while the sides of the canoe were about as watertight as a collander. Inevitably, by the time you reached the opposite bank you were sitting in about six inches of water. When we arrived with our equipment on the water-worn steps in the granite cliff that formed the landing stage, we found the ferry was at the opposite shore, so while Ben, Agustine, and the enormous African (whom we had christened Gargantua) lifted their voices and roared at the ferryman to return with all speed, Bob and I squatted in the shade and watched the usual crowd of Mamfe people bathing and washing in the brown waters below.

Swarms of small boys leaped shrieking off the cliffs and splashed into the water, and then shot to the surface again, their palms and the soles of their feet gleaming shell pink, their bodies like polished chocolate. The damsels, more demure, bathed with their sarongs on, only to emerge from the water with the cloth clinging to their bodies so tightly that it left nothing to the imagination. One small toddler, who could not have been more than five or six, made his way carefully down the cliff, his tongue protruding with concentration, carrying on his head an enormous water jar. On reaching the edge of the water he did not pause to remove the jar from his head, or to take off his sarong. He walked straight into the water and waded slowly and determinedly out into the river until he completely disappeared, and only the jar could be seen moving mysteriously along the surface of the water. At last this too disappeared. There was a moment's pause, and then the jar reappeared, this time moving shorewards, and eventually, beneath it, the boy's head bobbed up. He gave a tremendous snort to expel the air from his lungs, and then struggled grimly towards the beach, the now brimming jar on his head. When he reached the shore he edged the jar carefully onto a ledge of rock, and then re-entered the water, still wearing his sarong. From some intricate fold in this garment he produced a small fragment of Lifebuoy soap, and proceeded to rub it all over himself and the sarong with complete impartiality. Presently, when he had worked up such a lather all over himself that he looked like an animated pink snowman, he ducked beneath the surface to wash off the soap; then he waded ashore, settled the jar once more on his head, and slowly climbed the cliff and disappeared. It was the perfect example of the African application of the time-and-motion study.

By this time the ferry had arrived, and Ben and Agustine were arguing hotly with its aged occupant. They wanted him, instead of taking us straight across the river, to paddle us about half a mile upstream to a large sandbank. This would save us having to walk about a mile along the bank to reach the path that led to the forest. The old man appeared to be singularly obstinate about it all.

"What's the matter with him, Ben?" I inquired.

"Eh! Dis na foolish man, sah," said Ben, turning to me in exasperation. " 'E no agree for take us for up de river."

"Why you no agree, my friend?" I asked the old man. "If you go take us I go pay you more money and I go dash you."

"Masa," said the old man firmly, "dis na my boat, and if I go lose um I no fit catch money again. I no get chop for my belly. I no get one one penny."

"But how you go lose you boat?" I asked in amazement, for I knew that strip of river, and there were no rapids or bad currents along its length.

"Ipopo, Masa," explained the old man.

I stared at the ferryman, wondering what on earth he was talking about. Was Ipopo perhaps some powerful local *juju* I had not come across before?

"Dis Ipopo," I asked soothingly, "which side 'e live?"

"Wah! Masa never see um?" asked the old man in astonishment. " 'E dere dere for water close to D.O.'s house. 'E big like so-so motor. 'E de holla . . . 'e de get power too much."

"What's he talking about?" asked Bob, puzzled.

Suddenly, light dawned on me. "He's talking about the hippo herd in the river below the D.O.'s house, but it's such a novel abbreviation of the word that he had me foxed for a moment."

"Does he think they're dangerous?"

"Apparently, though I can't think why. They were perfectly placid last time I was here."

"Well, I hope they're still placid," said Bob.

I turned to the old man again. "Listen, my friend. If you go take us for up dis water, I go pay you six shilling and I go dash you cigarette, eh? And if sometime dis Ipopo go damage dis your boat I go pay for new one, you hear?"

"I hear, sah," said the old man, avarice struggling with caution.

"You agree?"

"I agree, sah."

We progressed slowly upstream, squatting in half an inch of water in the belly of the canoe.

"I suppose they can't really be dangerous," said Bob casually, trailing his hand nonchalantly in the water.

"When I was here last I used to go up to within thirty feet of them in a canoe and take photographs," I said.

"Dis Ipopo get strong head now, sah," said Ben tactlessly. "Two months pass dey kill three men and break two boats."

"That's a comforting thought," said Bob.

Ahead of us the brown waters were broken in many places by rocks. At any other time they would have looked exactly like rock, but now each one looked exactly like the head of a hippo, a cunning, maniacal hippo, lurking in the dark waters, awaiting our approach. Ben, presumably remembering his tale of daring with the bush cow, attempted to whistle, but it was a feeble effort, and I noticed that he scanned the waters ahead anxiously. After all, a hippo that has developed the habit of attacking canoes gets a taste for it, like a man-eating tiger, and will go out of his way to be unpleasant. They seem to regard it as a sport, and I was not feeling in the mood for gamboling in twenty feet of murky water with half a ton of sadistic hippo.

The old man, I noticed, was keeping our craft well into the bank, twisting and turning so that we were, as far as possible, always in shallow water. The cliff here was steep, but well supplied with footholds in case of emergency, for the rocks lay folded in great layers like untidy piles of fossilized magazines, overgrown with greenery. The trees that grew on top of the cliffs spread their branches well out over the water, so that we travelled in a series of fishlike jerks up a tunnel of shade, startling the occasional kingfisher that whizzed across our bows like a vivid blue shooting star, or a black and white wattled plover, that would flap away upstream, tittering imbecilely to itself, its feet grazing the water, its long yellow wattles flapping absurdly on each side of its beak.

Gradually we rounded the bend of the river, and there, some three hundred yards ahead, along the opposite shore, lay the white bulk of the sandbank, frilled with ripples. The old man gave a grunt of relief at the sight, and started to paddle more swiftly.

"Nearly there," I said gaily, "and not a hippo in sight."

It's always silly to make a remark like that. The words were hardly out of my mouth when a rock we were passing some fifteen feet away suddenly raised itself higher out of

the water and gazed at us with bulbous and astonished eyes, snorting out, at the same time, two slender fountains of spray, like a miniature whale.

Fortunately, our gallant crew resisted the impulse to leap out of the canoe *en masse* and swim for the bank. The old man drew in his breath with a sharp hiss, and dug his paddle deep into the water, so that the canoe pulled up short in a swirl and clop of bubbles. Then we sat and stared at the hippo, and the hippo sat and stared at us. Of the two, the hippo seemed the more astonished. Its chubby pink-grey face floated on the surface of the water like a disembodied head at a séance. Its great eyes stared at us with the innocent appraisal of a baby. Its ears flicked back and forth, as if it were waving to us. It sighed deeply, and moved a few feet nearer, still looking at us with wide-eyed innocence. Then, suddenly, Agustine let out a shrill whoop that made us all jump and nearly upset the canoe. We shushed him furiously, while the hippo continued its scrutiny of us unabashed.

"No de fear," said Agustine in a loud voice. "Na woman." He seized the paddle from the old man's reluctant grasp, and proceeded to beat on the water with the blade, sending up a shower of spray. The hippo opened its mouth in a gigantic yawn and displayed a length of tooth that had to be seen to be believed. Then, suddenly, and with apparently no muscular effort, its head sank beneath the surface. There was a moment's pause, during which we were all convinced that the beast was ploughing through the water somewhere directly beneath us, when its head rose to the surface again. This time, to our relief, it was some twenty yards up-river. It snorted out two more jets of spray, waggled its ears seductively, and sank again, only to reappear in a moment or so still farther up-stream. The old man grunted and retrieved his paddle from Agustine.

"Agustine, why you do dat foolish ting?" I asked in what I hoped was a steady and trenchant tone of voice.

"Sah, dat Ipopo no be man. Na woman dat," Agustine explained, hurt at my lack of faith in him.

"How you know?"

"Masa, I savvay all dis Ipopo for dis water," he explained. "Dis one na woman. Ef na man Ipopo 'e go chop us one time. But dis woman one no get strong head like 'e husband."

"Well, thank God for the weaker sex," I said to Bob, as the old man, galvanized into tremendous activity, sent the canoe shooting diagonally across the river, so that it ground onto the sandbank in a shower of pebbles. We unloaded our gear, told the old man to wait for us, and set off in the direction of the python's lair.

The path lay at first through some old native farmland, where the giant trees had been felled and now lay rotting across the ground. Between these trunks a crop of casava had been grown and harvested, and the ground allowed to lie fallow, so that the low growth of the forest—thorn bushes, convolvulus, and other tangled low growth—had swept into the clearing and covered everything with a cloak. There was always plenty of life to be seen in these abandoned farms, and as we pushed through the intricate web of undergrowth there were birds all around us. Beautiful little flycatchers hovered in the air, showing up powder-blue against the greenery; in the dim recesses of convolvulus-covered tree stumps robin-chats hopped perkily in search of grasshoppers, and looked startlingly like English robins; a pied crow flew up from the ground ahead and flapped heavily away, crying a harsh warning; in a thicket of thorn bushes, covered with pink flowers among which zoomed big blue bees, a kurrichane thrush gave us a waterfall of sweet song. The path wound its way through this moist, hot, waist-high undergrowth for some time, and then quite abruptly the undergrowth ended and the path led us out onto a golden grassfield, rippling with the heat haze.

Attractive though they were to look at, these fields were far from comfortable to walk across. The grass was tough and spiky, growing in tussocks carefully placed to trip the unwary traveller. In places, where sheets of grey rocks were exposed to the sun, the surface, sprinkled with a million tiny mica chips, sparkled and flashed in your eyes. The sun beat down upon your neck, and its reflections rebounded off the glittering surface of the rock and hit you in the face with the heat of a blast furnace. We plodded across this sun-drenched expanse, the sweat pouring off us.

"I hope this damned reptile's had the sense to go to ground where there's some shade," I said to Bob. "You could fry an egg on these rocks."

Agustine, who had been padding eagerly ahead, his sarong turning from scarlet to wine red as it absorbed the sweat from his body, turned and grinned at me, his face

freckled with a mass of sweat drops. "Masa hot?" he inquired anxiously.

"Yes, hot too much," I answered. " 'E far now dis place?"

"No, sah," he said pointing ahead. " 'E dere dere. Masa
never see dis man I done leave for watch?"

I followed his pointing finger and in the distance I could
see an area where the rocks had been pushed up and rum-
pled, like bedclothes, by some ancient volcanic upheaval,
so that they formed a miniature cliff running diagonally
across the grassfield. On top of this I could see the figures
of two more hunters, squatting patiently in the sun. When
they saw us they rose to their feet and waved ferocious-
looking spears in greeting.

" 'E dere dere for hole?" yelled Agustine anxiously.

" 'E dere dere," they called back.

When we reached the base of the small cliff, I could
quite see why the python had chosen this spot to stand
at bay. The rock face had been split into a series of shallow
caves, worn smooth by wind and water, each communi-
cating with the other, and the whole series sloping slightly
upwards into the cliff, so that anything that lived in
them would be in no danger of getting drowned in the
rainy season. The mouth of each cave was some eight feet
across and three feet high, which did not give anything
other than a snake much room for manœuvring. The hunters had very thoughtfully set fire to all the grass in the vicinity, in an effort to smoke the reptile out. The snake had
been unaffected by this, but now we had to work in a thick
layer of charcoal and feathery ash up to our ankles.

Bob and I got down on our stomachs and, shoulder to
shoulder, wormed our way into the mouth of the cave to
try to spot the python and map out a plan of campaign. We
soon found that the cave narrowed within three or four

feet of the entrance so that there was room for only one person, lying as flat as he could. After the glare of the sunshine outside the cave seemed twice as dark, and we could not see a thing. Our only indication that there was a snake there at all was a loud peevish hissing every time we moved. We called loudly for a torch, and when this had been unpacked and handed to us we directed its beam up the narrow passage.

Eight feet ahead of us the passage ended in a circular depression in the rock, and in this the python lay coiled, shining in the torchlight as if freshly polished. It was about fifteen feet long, as far as we could judge, and so fat that we pardoned Gargantua for comparing its girth with his enormous thigh. It was also in an extremely bad temper. The longer the torch beam played on it the more prolonged and shrill did its hisses become, until they rose to an eerie shriek. We crawled out into the sunlight again and sat up, both of us almost the same colour as our hunters under the thick layer of dark ash that had adhered lovingly to our sweaty bodies.

"The thing is to get a noose round its neck, and then we can all pull like hell and drag it out," said Bob.

"Yes, but the job's going to be to *get* the noose round its neck. I don't fancy being wedged in that passage if it decided to come down it after one. There's no room to manœuvre, and there's no room for anyone to help you if you do get entangled with it."

"Yes, that's a point." Bob admitted.

"There's only one thing to do," I decided. "Agustine, go quick-quick and cut one fork-stick for me. Big one. You hear?"

"Yes, sah," said Agustine. He whipped out his broad-bladed machete and trotted off towards the forest's edge some three hundred yards away.

"Remember," I warned Bob, "if we *do* succeed in yanking it out into the open you can't rely on the hunters. Everyone in the Cameroons is convinced that a python is poisonous; not only do they think its bite is deadly, but they also think it can poison you with the spurs under the tail. So if we do get it out it's no good grabbing the head and expecting them to hang on to the tail. You'll have to grab one end while I grab the other, and we'll just have to hope to heaven that they cooperate in the middle."

"That's a jolly thought," said Bob, sucking his teeth meditatively.

Presently Agustine returned, carrying a long, straight sapling with a fork at one end which he had cut. Onto this forked end I fastened a slip knot with some fine cord which, the manufacturers had assured me, would stand a strain of three hundredweight. Then I unravelled fifty feet or so of the cord, and handed the rest of the coil to Agustine.

"Now I go for inside, I go try put dis rope for 'e neck, eh? If I go catch 'e neck I go holla, and then all dis hunter man go pull one time. You hear?"

"I hear, sah."

"Now if I shout pull," I said, as I lowered myself delicately into the carpet of ash, "for Heaven's sake don't let them pull too hard. I don't want the damn thing pulled on top of me."

I wriggled slowly up the cave, carrying the sapling and cord with me, the torch in my mouth. The python hissed with undiminished ferocity and breath control. Then came the delicate job of trying to push the sapling ahead of me so that I could get the dangling noose over the snake's head. I found this impossible with the torch in my mouth, for at my slightest movement the beam swept everywhere but onto the point where I wanted it. I put the torch on

the ground, propped up on some rocks, its beam playing on the snake, and then, with infinite care, I edged the sapling up the cave towards the reptile. The python had, of course, coiled itself into a tight knot, with the head lying in the centre of the coils, so when I had got the sapling into position I had to get the snake to show its head. The only way I could do this was to prod the creature vigorously with the end of the sapling.

After the first prod the shining coils seemed to swell with rage, and there came echoing down the cave a hiss so shrill and so charged with malignancy that I almost dropped the sapling. Grasping the wood more firmly in my sweaty hand, I prodded again, and was treated to another shrill exhalation of breath. Five times I prodded before my efforts were rewarded. The python's head appeared suddenly over the top of the coils, and swept towards the end of the sapling, the mouth wide open and gleaming pinkly in the torchlight. But the movement was so sudden that I had no chance to get the noose over its head. The snake struck three times, and each time I made ineffectual attempts to noose it. My chief difficulty was that I could not get close enough, and so I was working at the full stretch of my arm; this, combined with the weight of the sapling, made my movements very clumsy. At last, dripping with sweat and with an aching arm, I crawled out into the sunlight.

"It's no good," I said to Bob. "It keeps its head buried in its coils and only pops it out to strike. You don't get a real chance to noose it."

"Let me have a go," he said eagerly.

He seized the sapling and crawled into the cave. There was a long pause during which all we could see was his large feet scrabbling and scraping for a foothold in the cave entrance. Presently he reappeared, cursing fluently.

"It's no good," he said. "We'll never get it with this."

"If they get us a forked stick like a shepherd's crook do you think you could get hold of a coil and pull it out?" I inquired.

"I think so," said Bob, "or at any rate I could probably make it uncoil so we can get a chance at the head."

So Agustine was once more dispatched to the forest with minute instructions as to the sort of stick we needed, and he soon returned with a twenty-foot branch at one end of which was a fishhook-like projection. "If you could crawl in with me and shine the torch over my shoulder it would help," said Bob. "If I put it on the ground I knock it over every time I move."

So we crawled into the cave together and lay there, wedged shoulder to shoulder. While I shone the torch down the tunnel Bob slowly edged his gigantic crook down towards the snake. Slowly, so as not to disturb the snake unnecessarily, he edged the hook over the top coil of the mound, settled it in place, shuffled his body into a more comfortable position, and then hauled with all his strength.

The results were immediate and confusing. To our surprise the entire bulk of the snake—after a momentary resistance—slid down the cave towards us. Exhilarated, Bob shuffled backwards (thus wedging us both more tightly in the tunnel) and hauled again. The snake slid still nearer and then started to unravel. Bob hauled again, and the snake uncoiled still further; its head and neck appeared out of the tangle and struck at us. Wedged like a couple of outsize sardines in an undersized can, we had no way to move except backwards, and so we slid backwards on our stomachs as rapidly as we could. At last, to our relief, we reached a slight widening in the passage, and this allowed us more room to manœuvre. Bob laid hold of the sapling and pulled at it grimly. He reminded me of a lanky and earnest blackbird tugging an outsize worm from its hole. The snake slid into view, hissing madly, its coils shuddering with muscular contraction as it tried to free itself of the hook round its body. Another good heave, I calculated, and Bob would have it at the mouth of the cave. I crawled out rapidly.

"Bring dat rope," I roared to the hunters. "Quick . . . quick . . . rope."

They leaped to obey as Bob appeared at the cave mouth; he scrambled to his feet and stepped back for the final jerk that would drag the snake out into the open where we could fall on it. But as he stepped back he put his foot on a loose rock which twisted under him, and he fell flat on his back. The sapling was jerked from his hands, the snake gave a mighty heave that freed its body from the hook, and with the smooth fluidity of water soaking into blotting paper it slid into a crack in the cave wall that did not look as though it could accommodate a mouse. As the last four feet of its length were disappearing into the bowels of the earth, Bob and I fell on it and hung on like grim death. We could feel the rippling of the powerful muscles as the snake, buried deep in the rocky cleft, struggled to break our grip on its tail. Slowly, inch by inch, the smooth scales slipped through our sweaty hands, and then, suddenly, the snake was gone. From somewhere deep in the rocks came a triumphant hiss.

Covered with ash and charcoal smears, our arms and legs scraped raw, our clothes black with sweat, Bob and I sat and glared at each other, panting for breath. We were so annoyed we were past speech.

"Ah, 'e done run, Masa," pointed out Agustine, who seemed to have a genius for underlining the obvious.

"Dat snake 'e get power too much," observed Gargantua moodily.

"No man fit hold dat snake for inside hole," said Agustine, attempting to comfort us.

" 'E get plenty, plenty power," intoned Gargantua again. " 'E get power pass man."

In silence I handed round the cigarettes, and we squatted in the carpet of ash and smoked.

"Well," I said at last, philosophically, "we did the best we could. Let's hope for better luck next time."

Bob, however, refused to be comforted. To have had the python of his dreams so close to capture and then to lose it was almost more than he could bear. He prowled around, muttering savagely to himself, as we packed up the nets and ropes, and then followed us moodily as we set off homewards.

The sun was now low in the sky, and by the time we had crossed the grassfield and entered the abandoned farmland a greenish twilight had settled on the world. Everywhere in the moist undergrowth giant glowworms gleamed and shuddered like sapphires, and through the warm air fireflies drifted, pulsating briefly like pink pearls against the dark undergrowth. The air was full of the evening scents, wood smoke, damp earth, the sweet smell of blossom already wet with dew. An owl called in an ancient, trembling voice, and another answered it.

The river was like a moving sheet of bronze in the twilight as we scrunched our way across the milk-white sandbank. The old man and the boy were curled up asleep in the bows of the canoe. They awoke, and, in silence, paddled us down the dark river. On the hilltop, high above us, we could see the lamps of the house shining out, and faintly, as a background to the swish and gurgle of our paddles, we could hear the gramophone playing, A drift of small white moths enveloped the canoe as it headed in towards the bank. The moon, very fragile and weak, was edging its way up through the filigree of the forest behind us, and once more the owls called, sadly, longingly, in the gloom of the trees.

MAIL BY HAND

To: Mr. G. Durrell,
The Zoological Department,
U.A.C. House,
Mamfe

Dear Sir,
    Here are two animals I am senting you like those animals
that you should me in the pictures. Any tipe of money you
want to sent to me try and rapp the money in a small piece
of paper and sent it to that boy that brought animals. You
know realy that a hunter always be derty so you should try
to send me one bar soap.
    Good greetings to you.

Yrs.
Peter N'amabong

2 : THE BALD-HEADED BIRDS

On the opposite bank of the Cross River, eight miles through the deep forest, lay the tiny village of Eshobi. I knew it and its inhabitants well, for on a previous trip I had made it one of my bases for a number of months. It had been a good hunting ground, and the Eshobi people had been good hunters, so, while we were in Mamfe, I was anxious to get in touch with the villagers and see if they could get us some specimens. As the best way of obtaining information, or of sending messages, was via the local market, I sent for Phillip, our cook. He was an engaging character, with a wide, buck-toothed smile, and a habit of walking with a stiff military gait and of standing at attention when spoken to that argued an army training, which, in fact, he had not had. He clumped up on to the veranda and stood before me as rigid as a guardsman.

"Phillip, I want to find an Eshobi man, you hear?" I said.

"Yes, sah."

"Now, when you go for market you go find me one

Eshobi man and you go bring him for here and I go give him book for take Eshobi, eh?"

"Yes, sah."

"Now, you no go forget, eh? You go find me Eshobi man one time."

"Yes, sah," said Phillip, and clumped off to the kitchen. He never wasted time on unnecessary conversation.

Two days passed without an Eshobi man putting in an appearance, and, occupied with other things, I forgot the whole matter. Then, on the fourth day, Phillip appeared, clumping down the drive triumphantly with a rather frightened-looking fourteen-year-old boy in tow. The lad had obviously clad himself in his best clothes for his visit to the metropolis of Mamfe, a fetching outfit that consisted of a tattered pair of khaki shorts and a grubby white shirt which had obviously been made out of a sack of some sort and had across its back the mysterious but decorative message PRODUCE OF GR in blue lettering. On his head was perched a straw hat which, with age and wear, had attained a pleasant shade of pale silvery green. This reluctant apparition was dragged up onto the front veranda, and his captor stood smugly to attention with the air of one who has, after much practice, accomplished a particularly difficult conjuring trick. Phillip had a curious way of speaking which had taken me some time to understand, for he spoke pidgin very fast and in a sort of muted roar, a cross between the sound of a bassoon and that of a regimental sergeant major, as though everyone in the world were deaf. When labouring under excitement he became almost incomprehensible.

"Who is this?" I asked, surveying the youth.

Phillip looked rather hurt. "Dis na man, sah," he roared, as if explaining something to a particularly dim-witted child. He gazed at his protégé with affection and gave the unfortunate lad a slap on the back that almost knocked him off the veranda.

"I can see it's a man," I said patiently, "but what does he want?"

Phillip frowned ferociously at the quivering youth and gave him another blow between the shoulder blades. "Speak now," he blared. "Speak now, Masa de wait."

We waited expectantly. The youth shuffled his feet, twiddled his toes in an excess of embarrassment, gave a shy watery smile, and stared at the ground. We waited patiently. Suddenly he looked up, removed his headgear, ducked his head, and said, "Good morning, sah," in a faint voice.

Phillip beamed at me as if this greeting had been sufficient explanation for the lad's presence. Deciding that my cook had not been designed by nature to play the part of a skilled and tactful interrogator, I took over myself.

"My friend," I said, "how dey de call you?"

"Peter, sah," he replied miserably.

"Dey de call um Peter, sah," bellowed Phillip, in case I should have been under any misapprehension.

"Well, Peter, why you come for see me?" I asked.

"Masa, dis man your cook 'e tell me Masa want some man for carry book to Eshobi," said the youth aggrievedly.

"Ah! You be Eshobi man?" I asked, light dawning.

"Yes, sah."

"Phillip," I said, "you are a congenital idiot."

"Yes, sah," said Phillip, pleased with this unsolicited testimonial.

"Why you never tell me dis be Eshobi man?"

"Wah!" gasped Phillip, shocked to the depths of his sergeant major's soul, "but I done tell Masa dis be man."

Giving Phillip up as a bad job, I turned back to the

youth. "Listen, my friend, you savvay for Eshobi one man dey de call Elias?"

"Yes, sah, I savvay um."

"All right. Now you go tell Elias dat I done come for Cameroon again for catch beef, eh? You go tell um I want um work hunter man again for me, eh? So you go tell um he go come for Mamfe for talk with me. You go tell um, say, dis Masa 'e live for U.A.C. Masa's house, you hear?"

"I hear, sah."

"Right, so you go walk quick-quick to Eshobi and tell Elias, eh? I go dash you dis cigarette so you get happy when you walk for bush."

He received the packet of cigarettes in his cupped hands, ducked his head, and beamed at me. "Tank you, Masa," he said.

"All right. Go for Eshobi now. Walker good."

"Tank you, Masa," he repeated, and stuffing the packet into the pocket of his unorthodox shirt he trotted off down the drive.

Twenty-four hours later Elias arrived. He had been one of my permanent hunters when I had been in Eshobi, and so I was delighted to see his fat, waddling form coming down the drive towards me, his Pithecanthropic features split into a wide grin of delighted recognition. Our greetings over, he solemnly handed me a dozen eggs carefully wrapped in banana leaves, and I reciprocated with a carton of cigarettes and a hunting knife I had brought out from England for that purpose. Then we got down to the serious business of talking about beef. First he told me about all the beef he had hunted and captured in my eight years' absence, and how my various hunter friends had got on. Old N'ago had been killed by a bush cow; Andraia had been bitten in the foot by a water beef; Samuel's gun had exploded and blown a large portion of his arm away (a good joke, this), while just recently John had killed the biggest bush pig they had ever seen, and sold the meat for over two pounds. Then, quite suddenly, Elias said something that riveted my attention.

"Masa remember dat bird Masa like too much?" he inquired in his husky voice.

"Which bird, Elias?"

"Dat bird 'e no get beer-beer for 'e head. Last time Masa live for Mamfe I done bring um two picken dis bird."

"Dat bird who make his house with potta-potta? Dat one who get red for his head?" I asked excitedly.

"Yes, na dis one," he agreed.

"Well, what about it?" I said, expectantly.

"When I hear Masa done come back for Cameroons I done go for bush for look dis bird," Elias explained. "I remember dat Masa 'e like dis bird *too much*. I look um, look um for bush for two, three days." He paused and looked at me, his eyes twinkling.

"Well?"

"I done find um, Masa," he said, grinning.

"You find um?" I could scarcely believe my luck. "Which side 'e dere? Which side 'e live? How many you see? What kind of place—"

" 'E dere dere," Elias went on, interrupting my flow of feverish questions, "for some place 'e get big big rock. 'E live for up hill, sah. 'E get 'e house for some big rock."

"How many house you see?"

"I see three, sah. But 'e never finish one house, sah."

"What's all the excitement about?" inquired Jacquie, who had just come out on to the veranda.

"Picathartes," I said succinctly, and it's to her credit that she knew exactly what I was talking about.

Picathartes was a bird that, until a few years ago, was known only from a few museum skins, and had probably been seen in the wild state by about two Europeans. Cecil Webb, then the London Zoo's official collector, managed to catch and bring back alive the first specimen of this extraordinary bird. Six months later, when I had been in the Cameroons, I had had two adult specimens brought in to me, but these had, unfortunately, died of aspergillosis, a particularly virulent lung disease, on the voyage home. Now Elias had found a nesting colony of them and it seemed, with luck, as though we might be able to get some fledglings and hand-rear them.

"Dis bird, 'e get picken for inside 'e house?" I asked Elias.

"Sometime 'e get, sah," he said doubtfully. "I never look for inside de house. I fear sometime de bird go run."

"Well," I said, turning to Jacquie, "there's only one thing to do, and that's to go to Eshobi and have a look. You and Sophie hang on here and look after the collection; I'll take Bob and spend a couple of days there after Picathartes. Even if they haven't got any young I would like to see the thing in its wild state."

"All right. When will you go?" asked Jacquie.

"Tomorrow, if I can arrange carriers. Give Bob a shout and tell him we're really going into the forest at last. Tell him to sort out his snake-catching equipment."

So, early the next morning, when the air was still comparatively cool, eight Africans appeared outside John Henderson's house, and after the usual bickering as to who should carry what they loaded our bundles of equipment onto their woolly heads and we set off for Eshobi. Having crossed the river, our little cavalcade made its way across the grassfield, where we had had our abortive python hunt, and on the opposite side we plunged into the mysteriousness of the forest. The Eshobi path lay twisting and turning through the trees in a series of intricate convolutions that would have horrified a Roman road builder. Sometimes it would double back on itself to avoid a huge rock, or a fallen tree, and at other times it would run, straight as a rod, through obstacles, so that our carriers were forced to stop and form a human chain to lift the loads over a tree trunk, or lower them down a small cliff.

I had warned Bob that we would see little, if any, wildlife on the way, but this did not prevent him from attacking every rotten tree trunk we passed, in the hope of unearthing some rare beast from its interior. I am tired of hearing and reading about the dangerous, evil tropical forest, teeming with wild beasts. In the first place it's about as dangerous as the New Forest in midsummer, and in the second place it does not teem with wild life, in the sense that every bush is *not* aquiver with some savage creature waiting to pounce. The animals are there, of course, but they very sensibly keep out of your way. I defy anyone to walk through the forest to Eshobi, and, at the end of it, be able to count on the fingers of both hands the "wild beasts" he has seen. How I wish that every bush did contain some "savage denizen of the forest" waiting to pounce; it would make the collector's job so much easier.

The only wild creatures at all common along the Eshobi path were butterflies, and these, obviously not having read the right books, showed a strong disinclination to attack us. Whenever the path dipped into a small valley, at the bottom would lie a tiny stream, and on the damp, shady banks alongside the clear waters the butterflies would be sitting in groups, their wings opening and closing slowly, so that from a distance some areas of the stream banks took on a opalescent quality, changing from flame red to white, from sky blue to mauve and purple, as the insects—in a sort of trance—seemed to be applauding the cool shade with their wings. The brown, muscular legs of the carriers would tramp through them unseeingly, and suddenly you would be waist-high in a swirling merry-go-round of colour as the butterflies dipped and wheeled around you; then, when you passed, they settled again on the dark soil which was as rich and moist as a fruit cake, and just as fragrant.

One vast and ancient tree marked the halfway point on the Eshobi road, a tree so tangled in a web of lianas as to be almost invisible. This was a resting place, and the carriers, grunting and exhaling their breath sharply through their front teeth in a sort of exhausted whistle, lowered their loads to the ground and squatted beside them, the sweat glistening on their bodies. I handed round cigarettes and we sat and enjoyed them quietly; in the dim, cathedral-like gloom of the forest there was no breeze, and the smoke rose in straight, swaying blue columns into the air. The only sounds were the incessant, circular-saw-like songs of the great green cicadas clinging to every tree, and, in the distance, the drunken honking of a flock of hornbills.

As we smoked we watched some of the little brown forest skinks hunting among the roots of the trees around us. These little lizards always looked neat and shining, as though they had been cast in chocolate and had just that second stepped out of the mould, gleaming and immaculate. They moved slowly and deliberately, as if they were afraid of getting their beautiful skins dirty. They peered from side to side with bright eyes as they slid through their world of brown, dead leaves, forests of tiny toadstools, and lawns of moss that padded the stones like a carpet. Their prey was the immense population of tiny creatures that inhabited the forest floor, the small black beetles hurrying along like undertakers late for a funeral, the slow, smooth-sliding slugs weaving a silver filigree of slime over the leaves, and the small, nut-brown crickets who squatted in the shadows waving their immensely long antennae to and fro, like amateur fishermen on the banks of a stream.

Among the dark, damp hollows between the buttress roots of the great tree under which we sat there were small clusters of an insect which had never failed to fascinate me. They looked like small daddy longlegs in repose, but with opaque, misty-white wings. They would sit there in groups of about ten, trembling their wings gently, and moving their fragile legs up and down, like restive horses. When disturbed they would all take to the air and start a combined operation which was quite extraordinary to watch. They would rise some eight inches into the air, form a circle in an area that could be covered by a saucer, and then start to fly round and round very rapidly, some going up and over, as it were, while the others swept round and round like a wheel. The effect from a distance was rather weird, for they resembled a whirling ball of shimmering misty white, changing its shape slightly at intervals, but always maintaining exactly the same position in the air. They flew so fast, and their bodies were so slender, that all you could see was this shimmer of frosty wings. This aerial display fascinated me so much that I used to go out of my way, when walking in the forest, to find groups of these insects and disturb them so that they would dance for me.

Eventually, we reached Eshobi at midday, and I found it had changed little from the days when I had been there eight years previously. There was still the same straggle of dusty thatched huts in two uneven rows, with a wide area of dusty path lying between them that acted as the village high street, a playground for children and dogs, and a scratching ground for the scrawny fowls. Elias came waddling down this path to greet us, picking his way carefully through the sprawling mass of babies and livestock, followed by a small boy carrying two large green coconuts on his head.

"Welcome, Masa, you done come?" he called huskily.

"*Iseeya*, Elias." I replied.

He grinned at us delightedly as the carriers, grunting and whistling, deposited our equipment all over the village street.

"Masa go drink dis coconut?" Elias asked hopefully, waving his machete about.

"Yes, we like um too much," I said, regarding the huge nuts thirstily.

Elias bustled into activity. From the nearest hut were brought two dilapidated chairs, and Bob and I were seated in a small patch of shade in the centre of the village street, surrounded by a crowd of politely silent but deeply fascinated Eshobites. With quick, accurate strokes of his machete Elias stripped away the thick husk from the coconut. When the tips of the nuts were exposed he gave each of them a swift slice with the end of his machete blade, and then handed them to us, each neatly trepanned so that we could drink the cool sweet juice inside. In each nut there was about two and a half glassfuls of this thirst-quenching, hygienically sealed nectar, and we savoured every mouthful.

Having rested, we next had to get the camp in order. Some two hundred yards from the village there was a small stream, and on its banks we chose an area that would not be too difficult to clear. A group of men armed with machetes set to work to cut down all the small bushes and saplings, while another group followed behind with short-handled, broad-bladed hoes, in an effort to level the red earth. At length, after the usual African uproar of insults, accusations of stupidity, sit-down strikes, and minor brawls, the area had been worked over so that it resembled a badly ploughed field, and we could get the tents up. While a meal was being prepared we went down to the stream and washed the dirt and sweat from our bodies in the icy water, watching the pink and brown crabs waving their pincers to us from among the rocks, and feeling the tiny brilliant blue and red fish nibbling gently at our feet. We wended our way back to camp, feeling refreshed, and found some sort of organization reigning. When we had eaten, Elias came and squatted in the shade of our lean-to tent, and we discussed hunting plans.

"What time we go look dis bird, Elias?"

"Eh, Masa savvay now 'e be hot too much. For dis time dis bird 'e go look for chop for bush. For evening time when it get cold 'e go for dis 'e house for work, and den we go see um."

"All right, then you go come back for four o'clock time, you hear? Then we go look dis bird, eh?"

"Yes sah," said Elias, rising to his feet.

"And if you no speak true, if we never see dis bird, if you've been funning me I go shoot you, bushman, you hear?"

"Eh!" he exclaimed, chuckling. "I never fun with Masa, for true, sah."

"All right, we go see you, eh?"

"Yes sah," he said, as he twisted his sarong round his ample hips and padded off towards the village.

At four o'clock the sun had dipped behind the tallest of the forest trees, and the air had the warm, drowsy stillness of evening. Elias returned, wearing, in place of his gaudy sarong, a small bit of dirty cloth twisted round his loins. He waved his machete nonchalantly.

"I done come Masa," he proclaimed. "Masa ready?"

"Yes," I said, shouldering my field glasses and collecting bag. "Let's go, hunter man."

Elias led us down the dusty main street of the village, and then branched off abruptly down a narrow alleyway between the huts. This led us into a small patch of farmland, full of feathery casava bushes and dusty banana plants. Presently the path dipped across a small stream and then wound its way into the forest. Before we had left the village street Elias had pointed out a hill to me which he said was the home of Picathartes, and although it had looked near enough to the village I knew better than to believe it. The Cameroon forest is like the "Looking-Glass" garden. Your objective seems to loom over you, but as you walk towards it it appears to shift position. At times, like Alice, you are forced to walk in the opposite direction in order to get there. And so it was with this hill. The path, instead of making straight for it, seemed to weave to and fro through the forest in the most haphazard fashion, until I began to feel I must have been looking at the wrong hill when Elias had pointed it out to me. At that moment, however, the path started to climb in a determined manner, and it was obvious that we had reached the base of the hill. Elias left the path and plunged into the undergrowth on one side, hacking his way through the overhanging lianas and thorn bushes with his machete, hissing softly through his teeth, his feet spreading out in the soft leaf mould without a sound. In a very short time we were plodding up a slope so steep that, on occasions, Elias's feet were on a level with my eyes.

The majority of hills and mountains in the Cameroons are of a curious and exhausting construction. Created by ancient volcanic eruption they were pushed skywards viciously by the massive underground forces, and this

shaped them in a peculiar way. They are curiously geo-
metrical in shape, some perfect isosceles triangles, some
acute angles, some cones, and some box-shaped. They
reared up in such a bewildering variety of shapes that it
would have been no surprise to see a cluster of them dem-
onstrating one of the more spiky and incomprehensible
of Pythagoras's theorems.

The hills whose sides we were now assaulting reared up
in an almost perfect cone. After you had been climbing for
a bit you began to gain the impression that it was much
steeper than it had first appeared, and within a quarter of
an hour you were convinced that the surface rose at the
rate of one in one. Elias went up it as though it were a level
macadam road, ducking and weaving skilfully between the
branches and overhanging undergrowth, while Bob and I,
sweating and panting, struggled along behind, sometimes
on all fours, in an effort to keep pace with him. Then, to
our relief, just below the crest of the hill, the ground flattened out into a wide ledge, and through the tangle of
trees we could see, ahead of us, a fifty-foot cliff of granite,
patched with ferns and begonias, with a tumbled mass of
giant, water-smoothed boulders at its base.

"Dis na de place, Masa," said Elias, stopping and lowering his fat bottom onto a rock.

"Good," said Bob and I in unison, and sat down to regain
our breath.

When we had rested Elias led us along through the maze
of boulders to a place where the cliff face sloped outwards,
overhanging the rocks below. We moved some little way
along under this overhang, and then Elias stopped suddenly.

"Dere de house, Masa," he said, his fine teeth gleaming
in a grin of pride. He was pointing up at the rock face, and
I saw, some ten feet above us, the nest of a picathartes.

At first glance it resembled a huge swallow's nest, made out of reddish-brown mud and tiny rootlets. At the base of the nest longer roots and grass stalks had been woven into the earth so that they hung down in a sort of beard; whether this was just untidy workmanship on the part of the bird, or whether it was done for motives of camouflage, was difficult to judge. Certainly the trailing beard of roots and grass did disguise the nest, for at first sight it resembled nothing more than a tussock of grass and mud that had become attached to the gnarled, water-ribbed surface of the cliff. The whole nest was about the size of a football and, placed where it was under the overhang of the cliff, it was nicely protected from any rain.

The first thing to do was to see if the nest contained anything. Luckily, growing opposite was a tall, slender sapling, and so we shinned up this in turn and peered into the interior of the nest; to our annoyance it was empty, though ready for the reception of eggs, for it had been lined with fine roots woven into a springy mat. We moved a little way along the cliff and soon came upon two more nests, one complete like the first one, and one half finished. But there was no sign of young or eggs.

"If we go hide, small time dat bird go come, sah," said Elias.

"Are you sure?" I asked.

"Yes sah, for true, sah."

"All right, we'll wait small time."

Elias took us to a place where a cave had been scooped out of the cliff, its mouth almost blocked by an enormous boulder, and we crouched down behind this natural screen. We had a clear view of the cliff face where the nests hung, while being ourselves in shadow and almost hidden by the wall of stone in front of us. So we settled down to wait.

The forest was getting gloomy now, for the sun was

well down. The sky through the tangle of leaves and lianas above our heads was green flecked with gold, like the flanks of an enormous dragon seen between the trees. Now the very special evening noises had started. In the distance we could hear the rhythmic crash of a troupe of mona monkeys on their way to bed, leaping from tree to tree, with a sound like great surf on a rocky shore, punctuated by occasional cries of "Oink, oink," from some member of the troupe. They passed somewhere below us along the base of the hill, but the undergrowth was too thick for us to see them. Following them came the usual retinue of hornbills, their wings making fantastically loud whooping noises as they flew from tree to tree. Two of them crashed into the branches above us and sat there silhouetted against the green sky, carrying on a long and complicated conversation, ducking and swaying their heads, great beaks gaping, whining and honking hysterically at each other. Their fantastic heads, with the great beaks and sausage-shaped casques lying on top, bobbing and mowing against the sky, looked like some weird devil masks from a Ceylonese dance.

The perpetual insect orchestra had increased a thousand-fold with the approach of darkness, and the valley below us seemed to vibrate with their song. Somewhere a tree frog started up, a long trilling note, followed by a pause, as though he were boring a hole through a tree with a miniature pneumatic drill and had to pause now and then to let it cool. Suddenly I heard a new noise. It was a sound I had never heard before, and I glanced inquiringly at Elias. He had stiffened, and was peering into the gloomy net of lianas and leaves around us.

"Na whatee dat?" I whispered.

"Na de bird, sah."

The first cry had been quite far down the hill, but now

came another cry, much closer. It was a curious noise which can be described, rather inadequately, as like the sudden sharp yap of a pekinese, only more flutelike and plaintive. Again it came, and again, but we could not see the bird, though we strained our eyes in the gloom.

"D'you think it's Picathartes?" whispered Bob.

"I don't know. It's a noise I haven't heard before."

There was a pause, and then suddenly the cry was repeated, very near now, and we lay unmoving behind our rock. Not far in front of where we lay there grew a thirty-foot sapling, bent under the weight of a liana as thick as a bell rope that hung in loops around it, its main stem hidden in the foliage of some nearby tree. While the rest of the area we could see was gloomy and ill-defined, this sapling—lovingly entwined by its killer liana—was lit by the last rays of the setting sun, so that the whole setting was rather like a meticulous painting. And, as though a curtain had gone up on this miniature stage, a real live picathartes suddenly appeared before us.

I say suddenly and I mean it. Animals and birds in a tropical forest generally approach so quietly that they appear before you unexpectedly, as if placed there by magic. The thick liana fell in a huge loop from the top of the sapling, and on this loop the bird materialized, swaying gently on its perch, its head cocked on one side as if it were listening. Seeing any wild animal in its natural surroundings is a thrill, but to watch something that you know to be a great rarity, something that you know has been seen by only a handful of people before you, gives the whole thing an added excitement and spice. So Bob and I lay there staring at the bird with the ardent, avid expressions of a couple of philatelists who have just discovered a penny black in a child's stamp album.

Picathartes was about the size of a jackdaw, but its body had the plump, sleek lines of a blackbird. Its legs were long and powerful, and its eyes large and obviously keen. Its breast was a delicate creamy buff and its back and long tail a beautiful slate grey, pale and powdery looking. The edge of the wing was black, and this acted as a dividing line that showed up wonderfully the breast and back colours. But it was the bird's head that caught your attention and held it. It was completely bare of feathers; the forehead and top of the head were a vivid sky blue, the back a bright rose-madder pink, while the sides of the head and the cheeks were black. Normally a bald-headed bird looks rather revolting, as if it were suffering from some unpleasant and incurable disease, but Picathartes looked splendid with its tricoloured head, as if it were wearing a crown.

After it had perched on the liana for a minute or so it flew down to the ground and proceeded to work its way to and fro among the rocks in a series of prodigious leaps, quite extraordinary to watch. They were not ordinary bird-like hops, for Picathartes was projected into the air as if its powerful legs were springs. It disappeared from view among the rocks, and we heard it call. It was answered almost at once from the top of the cliff, and looking up we could see another picathartes on a branch above us, peering down at the nests on the cliff face. Suddenly it spiraled downwards and alighted on the edge of one of the nests, paused a moment to look about, and then leaned forward to tidy up a hairlike rootlet that had become disarranged. Then it leaped into the air—there is no other way to describe it—and swooped down the hill into the gloomy forest. The other one emerged from among the rocks and flew after it, and in a short time we heard them calling to each other plaintively among the trees.

"Ah," said Elias, rising and stretching himself, " 'e done go."

" 'E no go come back?" I asked, pommeling my leg, which had gone to sleep.

"No, sah. 'E done go for inside bush, for some big stick where 'e go sleep. Tomorrow 'e go come back for work dis 'e house."

"Well, we might as well go back to Eshobi, then."

Our progress down the hill was a much speedier affair than our ascent had been. It was now so dark under the canopy of trees that we frequently missed our footing and slid for considerable distances on our backsides, clutch-

ing desperately at trees and roots as we passed in an effort
to slow down. Eventually we emerged in the Eshobi high
street, bruised, scratched, and covered with leaf mould. I
was filled with elation at having seen a live picathartes,
but at the same time depressed by the thought that we
could not hope to get any of the youngsters. It was obviously useless hanging around in Eshobi, so I decided that
we would set off again for Mamfe the next day, and try and
do a little collecting as we went through the forest. One
of the most successful ways of collecting animals in the
Cameroons is to smoke out hollow trees, and on our way to
Eshobi I had noticed several huge trees with hollow in-
teriors, which I thought might well repay investigation.

So, early the next morning, we packed up our equipment, and sent the carriers off with it. Then, accompanied
by Elias and three other Eshobi hunters, Bob and I followed at a more leisurely pace.

The first tree was some three miles into the forest, lying
fairly close to the edge of the Eshobi road. It was some
hundred and fifty feet high, and the greater part of its
trunk was as hollow as a drum. There is quite an art to
smoking out a hollow tree. It is a prolonged and sometimes
complicated process. Before going to all the trouble of
smoking a tree, the first thing to do, if possible, is to ascertain whether or not there is anything inside worth smoking out. If the tree has a large hole at the base of the trunk,
as most of them do, this is a relatively simple matter. You
simply stick your head inside and get somebody to beat the
trunk with a stick. If there are any animals inside, you will
hear them moving about uneasily after the reverberations
have died away, and even if you can't hear them you can be
assured of their presence by the shower of powdery rotten wood that will come cascading down the trunk. Having
discovered that there is something inside the tree, you must

next scan the top part of the trunk with field glasses and try to spot all exit holes, for they have to be covered with nets. When this has been done, and a man stationed up the tree to retrieve any creature that gets caught up there, you then take the same precaution with the holes at the base of the trunk. Then you light your fire, and this is the really tricky part of the operation, for the interior of these trees is generally dry and tinder-like, and if you are not careful you can set the whole thing ablaze. So first of all you kindle a small bright blaze with dry twigs, moss, and leaves, and when this is well alight you carefully cover it with ever-increasing quantities of green leaves, so that the fire no longer blazes but sends up a sullen column of pungent smoke, which is sucked up the hollow barrel of the tree exactly as if it were a chimney. After this anything can happen, for these hollow trees can contain a weird variety of inhabitants, ranging from spitting cobras to civet cats, from giant snails to bats; so half the charm and excitement of smoking out a tree is that you are never quite sure what is going to appear.

The first tree we smoked was not a wild success. All we got was a handful of leaf-nosed bats with extraordinary gargoyle-like faces, three giant millipedes that looked like frankfurter sausages with a fringe of legs underneath, and a small grey dormouse which bit one of the hunters in the thumb and escaped. So we removed the nets, put out the fire, and proceeded on our way. The next hollow tree was considerably taller and of tremendous girth. At its base was an enormous split in the trunk shaped like a church door, and four of us could stand comfortably in the gloomy interior. Peering up the hollow barrel of the trunk and beating on the wood with a machete, we were rewarded by vague scuffling noises from above, and a shower of powdery rotten wood fell on our upturned faces and into our eyes. Obviously the tree contained something. Our chief problem was to get a hunter to the top of the tree to cover the exit holes, for the trunk swept up about a hundred and twenty feet into the sky as smooth as a walking stick. Eventually, we joined all three of our rope ladders together and tied a strong, light rope to one end. Then, after weighting the rope end, we hurled it up into the forest canopy until our arms ached, until at last it fell over a branch and we could haul the ladders up into the sky and secure them. So, when the nets were fixed in position at the top and bottom of the tree, we lit the fire at the base of the trunk and stood back to await results.

Generally one had to wait four or five minutes for the smoke to percolate to every part of the tree before there was any response, but in this particular case the results were almost immediate. The first beasts to appear were those nauseating-looking creatures called whip scorpions. They cover, with their long angular legs, the circumference of a soup plate, and they look like a nightmare spider that has been run over by a steamroller and reduced to a paper-like thinness. This enables them to slide in and out of crevices that would allow access to no other beast, in a most unnerving manner. Apart from this, they could glide about over the surface of the wood as though it were ice, and at a speed that was quite incredible. It was this speedy and silent movement, combined with such a forest of legs, that made them so repulsive, and made one instinctively shy away from them, even though we knew they were harmless. So, when the first one appeared magically out of a crack and scuttled over my bare arm as I leaned against the tree, it produced an extraordinarily demoralizing effect, to say the least. I had only just recovered from this when all the other inhabitants of the tree started to vacate in a body. Five fat grey bats flapped out into the nets, where they hung chittering madly and screwing up their faces in rage. They were quickly joined by two green forest squirrels, with pale fawn rings round their eyes, which uttered shrill grunts of rage as they rolled about in the meshes of the nets, and we tried to disentangle them without getting bitten. They were followed by six grey dormice, two large green rats with orange noses and behinds, and a slender green tree snake with enormous eyes, which slid calmly through the meshes of the nets with a slightly affronted air, and disappeared into the undergrowth before anyone could do anything sensible about catching it. The noise and confusion were incredible. Africans danced about through the billowing smoke, shouting instructions of which nobody took the slightest notice, getting bitten with shrill yells of agony, stepping on one another's feet, wielding machetes and sticks with gay abandon and complete disregard for safety. The man posted in the top of the tree was having fun on his own, and was shouting and yelling and leaping about in the branches with such vigour that I expected to see him crash to the forest floor at any moment. Our eyes streamed, our lungs were filled with smoke, but the collecting bags filled up with a wiggling, jumping cargo of creatures.

Eventually the last of the tree's inhabitants had appeared, the smoke had died down, and we could pause to smoke a cigarette and examine each other's honourable wounds. As we were doing this the man at the top of the tree lowered down two collecting bags on the end of long strings, before preparing to return to earth himself. I took the bags gingerly, not knowing what the contents were, and inquired of the stalwart at the top of the tree how he had got on.

"What you get for dis bag?" I asked.

"Beef, Masa," he replied intelligently.

"I know it's beef, bushman, but what kind of beef you get?"

"Eh! I no savvay how Masa call um. 'E so so rat, but 'e get wing. Dere be one beef for inside 'e get eye big big like man, sah."

I was suddenly filled with an inner excitement. "'E get hand like rat or like monkey?" I shouted.

"Like monkey, sah."

"What is it?" asked Bob interestedly, as I fumbled with the string round the neck of the bags.

"I'm not sure, but I think it's a bushbaby. If it is it can only be one of two kinds, and both of them are rare."

I got the string off the neck of the bag after what seemed an interminable struggle, and cautiously opened it. Regarding me from the interior was a small, neat grey face with huge ears folded back like fans against the side of the head, and two enormous golden eyes, that regarded me with the horror-stricken expression of an elderly spinster

in a bath who had discovered a man in the bathroom cup-
board. The creature had large, human-looking hands,
with long, slender, bony fingers. Each of these, except the
forefinger, was tipped with a small, flat nail that looked as
though it had been delicately manicured, while the forefinger possessed a curved claw that looked thoroughly out
of place on such a human hand.

"What is it?" asked Bob in hushed tones, seeing that I
was gazing at the creature with an expression of bliss on
my face.

"This," I said ecstatically, "is a beast I have tried to get
every time I've been to the Cameroons. *Euoticus elegan-
tulus,* better known to its friends as a needle-clawed lemur
or bushbaby. They're extremely rare, and if we succeed
in getting this one to England it will be the first ever to be
brought back to Europe."

"Gosh," said Bob, suitably impressed.

I showed the little beast to Elias. "You savvay dis beef,
Elias?"

"Yes, sah, I savvay um."

"Dis kind of beef I want *too much.* If you go get me more
I go pay you one one pound. You hear."

"I hear, sah. But Masa savvay dis kind of beef 'e come
out for night time. For dis kind of beef you go look um
with hunter light."

"Yes, but you tell all people of Eshobi I go pay one one
pound for dis beef, you hear?"

"Yes, sah. I go tell um."

"And now," I said to Bob, carefully tying up the bag
with the precious beef inside, "let's get back to Mamfe
quick and get this into a decent cage where we can see it."

So we packed up the equipment and set off at a brisk
pace through the forest towards Mamfe, pausing fre-

quently to open the bag and make sure that the precious specimen had enough air, and had not been spirited away by some frightful *juju*. We reached Mamfe at lunchtime and burst into the house, calling to Jacquie and Sophie to come and see our prize. I opened the bag cautiously and euoticus edged its head out and surveyed us all in turn with its enormous, staring eyes.

"Oh, isn't it *sweet*," said Jacquie.

"Isn't it a *dear*," crooned Sophie.

"Yes," I said proudly. "It's a—"

"What shall we call it?" asked Jacquie.

"We'll have to think of a good name for it," said Sophie.

"It's an extremely rare—" I began.

"How about Bubbles?" suggested Sophie.

"No, it doesn't look like a Bubbles," said Jacquie, surveying it critically.

"It's an euoticus. . . ."

"How about Moony?"

"No one has ever taken it back. . . ."

"No, doesn't look like a Moony either."

"No European zoo has ever . . ."

"What about Fluffykins?" asked Sophie.

I shuddered.

"If you must give it a name, call it Bug-Eyes," I said.

"Oh, yes!" said Jacquie. "That suits it."

"Good," I said. "I am relieved to know that we have successfully christened it. Now what about a cage for it?"

"Oh, we've got one here," said Jacquie. "Don't worry about that."

We eased the animal into the cage, and it squatted on the floor, glaring at us with unabated horror.

"Isn't it sweet?" Jacquie repeated.

"Is 'oo a poppet?" gurgled Sophie.

I sighed. It seemed that, in spite of all my careful training, both my wife and secretary relapsed into the most revolting fubsy attitude when faced with anything fluffy.

"Well," I said resignedly, "supposing you feed 'oos poppet? This poppet's going inside to get an itsy-bitsy slug of gin."

PART TWO: BACK TO BAFUT

MAIL BY HAND

My good friend,
I am glad that you have arrive once more to Bafut. I welcome you. When you are calm from your journeys come and see me.

Your good friend,
Fon of Bafut

3: THE FON'S BEEF

On our return from Eshobi, Jacquie and I loaded up our
truck with the cages of animals we had obtained to date,
and set out for Bafut, leaving Bob and Sophie in Mamfe
for a little longer to try to obtain some more of the rain-
forest animals.

The journey from Mamfe to the highlands was long
and tedious, but never failed to fascinate me. To begin
with, the road ran through thick forest, through the valley in which Mamfe lay. The truck roared and bumped
its way along the red road between gigantic trees, each
festooned with creepers and lianas, through which flew
small flocks of hornbills, honking wildly, or pairs of touracos, jade green with magenta wings flashing as they flew.
On the dead trees by the side of the road the lizards,
orange, blue, and black, vied with the pygmy kingfishers
over the spiders, locusts, and other succulent titbits to be
found amongst the purple and white convolvulus flowers.
At the bottom of each tiny valley would be a small stream,

spanned by a creaking wooden bridge, and as the truck roared across, great clouds of butterflies would rise from the damp earth at the sides of the water and swirl briefly round the hood. After a couple of hours the road started to climb, at first almost imperceptibly, in a series of great swinging loops through the forest, and here and there by the side of the road you could see the giant tree ferns like green fountains spouting miraculously out of the low growth. As one climbed higher, the forest gave way here and there to patches of grassland, bleached white by the sun.

Then, gradually, as though we were shedding a thick green coat, the forests started to drop away and the grassland took its place. The gay lizards ran sun-drunk across the road, and flocks of minute finches burst from the undergrowth and drifted across in front of us, their crimson feathering making them look like showers of sparks from some gigantic bonfire. The truck roared and shuddered, steam blowing up from the radiator, as it made the final violent effort and reached the top of the escarpment. Behind lay the Mamfe forest, in a million shades of green, and before us was the grassland, hundreds of miles of rolling mountains, lying in folds to the farthest dim horizons, gold and green, stroked by cloud shadows, remote and beautiful in the sun. The driver eased the truck on to the top of the hill and brought it to a shuddering halt that made the red dust swirl up in a waterspout that enveloped us and our belongings. He smiled the wide, happy smile of a man who has accomplished something worth doing.

"Why we stop?" I inquired.

"I go piss," explained the driver frankly, and he disappeared into the long grass at the side of the road.

Jacquie and I uncoiled ourselves from the red-hot interior of the cab and walked round to the back of the truck to see how our creatures were faring. Phillip, seated stiff and upright on a tarpaulin, turned to us a face made bright red with dust. His trilby, which had been a very delicate pearl grey when we started, was also bright red. He sneezed violently into a green handkerchief, and surveyed me reproachfully.

"Dust *too much*, sah," he roared at me, in case the fact had escaped my observation. As Jacquie and I were almost as dusty in the front of the truck, I was not inclined to be sympathetic.

"How are the animals?" I asked.

" 'E well sah. But dis bushhog, sah, 'e get strong head *too much*."

"Why, what the matter with it?"

" 'E done tief dis ma pillow," said Phillip indignantly.

I peered into the black-footed mongoose's cage. Ticky had whiled away the tedium of the journey by pushing her paw through the bars and gradually dragging in with her the small pillow which was part of our noble cook's bedding. She was sitting on the remains, looking very smug and pleased with herself, surrounded by snowdrifts of feathers.

"Never mind," I said consolingly. "I'll buy you a new one. But you go watch your other things, eh? Sometime she go tief them as well."

"Yes, sah, I go watch um," said Phillip, casting a black look at the feather-encrusted Ticky.

We drove on through the green, gold, and white grassland, under a blue sky veined with fine wisps of windwoven white cloud, like frail twists of sheep's wool blowing across the sky. Everything in this landscape seemed to be the work of the wind. The great outcrops of grey rocks were carved and ribbed by it into fantastic shapes; the long grass was curved over into frozen waves by it; the small trees had been bent, carunculated, and distorted by it. And the whole landscape throbbed and sang with the wind, hissing softly in the grass, making the small trees creak and whine, hooting and blaring round the towering cornices of rock.

Towards the end of the day the sky became pale gold. Then, as the sun sank behind the farthest rim of mountains, the world was enveloped in the cool green twilight, and in the dusk the truck roared round the last bend and drew up at the hub of Bafut, the compound of the Fon. To the left lay the vast courtyard, and behind it the clusters of huts in which lived the Fon's wives and children. Dominating them all was the great hut in which dwelt the spirit of his father, and a great many other lesser spirits, looming like a monstrous time-blackened beehive against the jade night sky. To the right of the road, perched on top of a tall bank, was the Fon's rest house, like a two-story Italian villa, stone-built and with a neatly tiled roof. It was shoebox-shaped, and both lower and upper stories were surrounded by wide verandas, festooned with bougain-villaea covered with pink and brick-red flowers.

Wearily we climbed out of the truck and supervised the unloading of the animals and their installation on the top-story veranda. Then the rest of the equipment was un-loaded and stored, and while we made vague attempts to wash some of the red dust off our bodies, Phillip seized the remains of his bedding and his box full of cooking utensils and food and marched off to the kitchen quarters in a stiff, brisk way, like a military patrol going to quell a small but irritating insurrection. By the time we had fed the ani-mals he had reappeared with an astonishingly good meal, and having eaten it we fell into bed and slept like the dead.

The next morning, in the cool dawn light, we went to pay our respects to our host, the Fon. We made our way across the great courtyard and plunged into the maze of tiny squares and alleyways formed by the huts of the Fon's wives. Presently we found ourselves in a small courtyard shaded by an immense guava tree, and there was the Fon's own villa, small, neat, stone- and tile-built, with a wide veranda along one side. And there, at the top of the steps that ran up to the veranda, stood my friend the Fon of Bafut.

He stood there, tall and slender, wearing a plain white robe embroidered with blue. On his head was a small skull-cap in the same colours. His face was split by the joyous, mischievous grin I knew so well, and he was holding out one enormous slender hand in greeting.

"My friend, *iseeya*," I called, hurrying up the stairs to him.

"Welcome, welcome. You done come. Welcome," he exclaimed, seizing my hand in his huge palm and draping a long arm round my shoulders and patting me affectionately.

"You well, my friend?" I asked, peering up into his face.

"I well, I well," he said, grinning.

It seemed to me to be an understatement: he looked positively blooming. He had been well into his seventies when I had last met him, eight years before, and he appeared to have weathered the intervening years better than I had. I introduced Jacquie, and was quietly amused by the contrast. The Fon, six feet three inches, and appearing taller because of his robes, towered beamingly over Jacquie's five feet one inch, and her hand was as lost as a child's would have been in the depths of his great dusky paw.

"Come, we go for inside," he said, and holding our hands led us into his villa.

The interior was as I remembered it, a cool, pleasant room with leopard skins on the floor, and wooden sofas, beautifully carved, piled high with cushions. We sat down, and one of the Fon's wives came forward carrying a tray with glasses and drinks on it. The Fon splashed scotch into three glasses with a liberal hand, and passed them round, beaming at us. I surveyed the four inches of neat spirit in the bottom of my glass and sighed. I could see that the Fon had not, in my absence, joined the temperance movement, whatever else he had done.

"Chirri-ho!" said the Fon, and downed half the contents of his glass at a gulp. Jacquie and I sipped ours more sedately.

"My friend," I said, "I happy too much I see you again."

"Wah! Happy?" said the Fon. "I get happy for see you. When dey done tell me you come for Cameroon again I get happy too much."

I sipped my drink cautiously. "Some man done tell me that you get angry for me because I done write dat book about dis happy time we done have together before. So I de fear for come back to Bafut," I said.

The Fon scowled. "Which kind of man tell you dis ting?" he inquired furiously.

"Some European done tell me."

"Ah! European," said the Fon, shrugging, as if surprised that I should believe anything told to me by a white person. "Na lies dis."

"Good," I said, greatly relieved. "If I think you get angry for me, my heart no go be happy."

"No, no, I no get angry for you," said the Fon, splashing another large measure of scotch into my glass before I could stop him. "Dis book you done write . . . I like um

foine. You done make my name go for all de world. Every
kind of people 'e know my name. Na foine ting dis."

Once again I realized I had underestimated the Fon's
abilities. He had obviously realized that any publicity is
better than none. "Look um," he went on, "plenty plenty
people come here for Bafut, all different different people,
dey all show me dis your book 'e get my name for inside.
Na foine ting dis."

"Yes, na fine thing," I agreed, rather shaken. I had had
no idea that I had unwittingly turned the Fon into a sort
of literary lion.

"Dat time I done go for Nigeria," he said, holding the
bottle of scotch up to the light pensively. "Dat time I done
go for Lagos to meet dat Queen woman, all dis European
dere 'e get dis your book. Plenty, plenty people dey ask
me for write dis ma name for inside dis your book."

I gazed at him open-mouthed; the idea of the Fon in
Lagos sitting and autographing copies of my book ren-
dered me speechless.

"Did you like the Queen?" asked Jacquie.

"Wah! Like? I like um too much. Na foine woman
dat. Na small, small woman, same same for you. But 'e
get power, time no dere. Wah! Dat woman get power
plenty."

"Did you like Nigeria?" I asked.

"I no like," said the Fon firmly. " 'E hot too much. Sun,
sun, sun, I shweat, I shweat. But dis Queen woman she get
plenty power. She walker walker she never shweat. Na
foine woman dis."

He chuckled reminiscently, and absent-mindedly
poured us all out another drink.

"I done give dis Queen," he went on, "dis teeth for elephant. You savvay um?"

"Yes, I savvay um," I said, remembering the magnif-

icent carved tusk the Cameroons had presented to Her
Majesty. "I done give dis teeth for all dis people of Cameroon," he explained. "Dis Queen she sit for some chair
an' I go softly softly for give her dis teeth. She take um.
Den all dis European dere day say it no be good ting for
show your arse for dis Queen woman, so all de people
walker walker backwards. I walker walker backwards.
Wah! Na step dere, eh? I de fear I de fall, but I walker
walker softly and I never fall . . . but I de fear too
much."

He chuckled over the memory of himself backing down
the steps in front of the Queen until his eyes filled with
tears.

"Nigeria no be good place," he said, "hot too much. I
shweat."

At the mention of sweat I saw his eyes fasten pensively
on the whisky bottle, so I rose hurriedly to my feet and
said that we really ought to be going, as we had a lot of un-
packing to do.

The Fon walked out into the sunlit courtyard with us,
and, holding our hands, peered earnestly down into our
faces. "For evening time you go come back," he said, "we
go drink, eh?"

"Yes, for evening time we go come," I assured him.

He beamed down at Jacquie. "For evening time I go
show you what kind of happy time we get for Bafut," he
said.

"Good," said Jacquie, smiling bravely.

The Fon waved his hands in elegant dismissal, and then
turned and made his way back into his villa, while we
trudged our way back to the rest house.

"I don't think I could face any breakfast after that
scotch," said Jacquie.

"But that wasn't drinking," I protested. "That was just

a sort of mild *apéritif* to start the day. You wait until tonight."

"Tonight I shan't drink. I'll leave it to you two," said
Jacquie firmly. "I shall have one drink and that's all."

After breakfast, while we were attending to the animals,
I happened to glance over the veranda rail and noticed on
the road below a small group of men approaching the
house. When they drew nearer I saw that each of them
was carrying either a raffia basket or a calabash with the
neck stuffed with green leaves. I could hardly believe that
they were bringing animals as soon as this, for generally it
takes up to a week for the news to get around and for the
hunters to start bringing stuff in. But as I watched them
with bated breath they turned off the road and started to
climb the long flight of steps up to the veranda, chattering
and laughing among themselves. Then, when they
reached the top step, they fell silent, and carefully laid
their offerings on the ground.

"*Iseeya,* my friends." I said.

"Morning, Masa," they chorused, grinning.

"Na whatee all dis ting?"

"Na beef, sah," they chorused.

"But how you savvay dat I done come for Bafut for buy
beef?" I asked, greatly puzzled.

"Eh, Masa, de Fon 'e done tell us," said one of the hunters.

"Good Lord, if the Fon's been spreading the news before
we arrived we'll be inundated in next to no time," said
Jacquie.

"We're pretty well inundated now," I said, surveying the
group of containers at my feet, "and we haven't even unpacked the cages yet. Oh well, I suppose we'll manage.
Let's see what they've got."

I bent down, picked up a raffia bag, and held it aloft.

"Which man bring dis?" I asked.

"Na me, sah."

"Na whatee dere for inside?"

"Na squill-lill, sah."

"What," inquired Jacquie, as I started to unravel the strings on the bag, "is a squill-lill?"

"I haven't the faintest idea," I replied.

"Well, hadn't you better ask?" suggested Jacquie practically, "for all you know it might be a cobra or something."

"Yes, that's a point." I turned to the hunter who was watching me anxiously. "Na whatee dis beef squill-lill?"

"Na small beef, sah."

"Na bad beef? 'E go chop man?"

"No, sah, at all. Dis one na squill-lill small, sah . . . na picken."

Fortified with this knowledge I opened the bag and peered into its depths. At the bottom, squirming and twitching in a nest of grass, lay a tiny squirrel some three and a half inches long. It couldn't have been more than a few days old, for it was still covered in the neat, shining plushlike fur of an infant, and it was still blind. I lifted it out carefully and it lay in my hand making faint squeaking noises like something out of a Christmas cracker, its pink mouth open in an O like a choirboy's, its minute paws making paddling motions against my fingers. I waited patiently for the flood of anthropomorphism to die down in my wife.

"Well," I said, "if you want it keep it. But I warn you it will be hell to feed. The only reason I can see for trying is because it's a baby black-eared, and they're quite rare."

"Oh, it'll be all right," said Jacquie optimistically. "It's strong, and that's half the battle."

I sighed. I remembered the innumerable baby squirrels

I had struggled with in various parts of the world, and how each one had seemed more imbecilic and more bent on self-destruction than the last. I turned to the hunter. "Dis beef, my friend. Na fine beef dis, I like um too much. But 'e be picken, eh? Sometime 'e go die-o, eh?"

"Yes, sah," agreed the hunter gloomily.

"So I go pay you two-two shillings now, and I go give you book. You go come back for two week time, eh, and if dis picken 'e alive I go pay you five-five shilling more, eh? You agree?"

"Yes sah, I agree," said the hunter, grinning delightedly.

I paid him the two shillings, and then wrote out a promissory note for the other five shillings, and watched him tuck it carefully into a fold of his sarong. "You no go lose um," I said, "if you go lose um I no go pay you."

"No, Masa, I no go lose um," he assured me, grinning.

"You know, it's the most beautiful colour," said Jacquie, peering at the squirrel in her cupped hands.

On that point I could agree with her. The diminutive head was bright orange, with a neat black rim behind each ear, as though its mother had not washed it properly. Its body was brindled green on the back and pale yellow on the tummy, while the ridiculous tail was darkish green above and flame orange below.

"What shall I call it?" asked Jacquie.

I glanced at the quivering scrap, still doing choral practice in her palm. "Call it what the hunter called it— Squill-lill Small," I suggested, and so Squill-lill Small she became, later to be abbreviated to Small for convenience.

While indulging in this problem of nomenclature I had been busy untying another raffia basket, without having taken the precaution of asking the hunter what it contained. So, when I opened it, a small, pointed, ratlike face appeared, bit me sharply on the finger, uttered a piercing shriek of rage, and disappeared into the depths of the basket again.

"What on earth was that?" asked Jacquie, as I sucked my finger and cursed, while all the hunters chorused, "Sorry, sah, sorry, sah," as though they had been collectively responsible for my stupidity.

"That fiendish little darling is a pygmy mongoose," I said. "For their size they're probably the fiercest creatures in Bafut, and they've got the most penetrating scream of any small animal I know, except a marmoset."

"What are we going to keep it in?"

"We'll have to unpack some cages. I'll leave it in the bag until I've dealt with the rest of the stuff," I said, carefully tying the bag up again.

"It's nice to have two different species of mongoose," said Jacquie.

"Yes," I agreed, sucking my finger. "Delightful."

The rest of the containers, when examined, yielded nothing more exciting than three common toads, a small green leaf viper, and four weaver birds which I did not want. Having disposed of them and the hunters, I turned my attention to the task of housing the pygmy mongoose. One of the worst things you can do on a collecting trip is to be unprepared in your caging. I had been caught like this on my first expedition, for, although we had taken a lot of various equipment, I had failed to take any ready-made cages, thinking there would be plenty of time to build them on the spot. The result was that the first flood of animals caught us unprepared and by the time we had struggled, night and day, to house them all adequately the second wave of creatures had arrived, and we were back where we started. At one point I had as many as six differ-

ent creatures tied to my camp bed on strings. After this experience, I have always taken the precaution of taking
some collapsible cages with me on a trip so that, whatever
else happens, I am certain that I can accommodate at least
the first forty or fifty specimens.

So I erected one of our specially built cages, filled it with
dry banana leaves, and then eased the pygmy mongoose
into it without getting bitten. It stood in the centre of the
cage, regarding me with small, bright eyes, one dainty
paw held up, and proceeded to utter shriek upon shriek of
fury until our ears throbbed. The noise was so penetrating
it was really painful, and so, in desperation, I threw a large
lump of meat into the cage. The pygmy leaped on it, shook
it vigorously to make sure it was dead, and then carried
it off triumphantly to a corner, where it settled down to
eat. It still continued to shriek at us, but the sounds were
now mercifully muffled by the food. I placed the cage next
to the one occupied by Ticky, the black-footed mongoose,
and sat down to watch.

At a casual glance you would not think that the two animals were even remotely related. The black-footed mongoose, although still only a baby, measured some two feet
in length and stood about eight inches in height. She had
a blunt, rather doglike face with dark, round, and somewhat protuberant eyes. Her body, head, and tail were a
rich, creamy white, while her slender legs were a rich
brown that was almost black. She was sleek, sinuous, and
svelte, and reminded me of a creamy-skinned Parisienne
belle-amie clad in nothing more than two pairs of black
silk stockings. In contrast, the pygmy mongoose looked
anything but Parisienne. It measured, including tail, some
ten inches in length. It had a tiny, sharply pointed face
with a small, circular pink nose and a pair of small, glitter-

ing, sherry-coloured eyes. Its fur, which was rather long and thick, was a deep chocolate brown with faint suggestions of ginger tinges in it here and there.

Ticky, who was very much the *grande dame*, peered out of her cage at the newcomer with something akin to horror on her face, watching fascinated as it shrieked and grumbled over its gory hunk of meat. Ticky was a very dainty and fastidious feeder herself and would never have dreamed of behaving in this uncouth way, yelling and screaming with your mouth full and generally carrying on as though you had never had a square meal in your life. She watched the pygmy for a moment or so, gave a sniff of scorn, turned round and round elegantly two or three times, and then lay down and went to sleep. The pygmy, undeterred by this comment on its behaviour, continued to champ and shrill over the last bloody remnants of its food. When the last morsel had been gulped down, and the ground around carefully inspected for any bits that might have been overlooked, it sat down and scratched itself vigorously for some minutes and then curled up and went to sleep as well. When we woke it up again about an hour later in order to record its voice for posterity, it produced such screams of rage and indignation that we were forced to move the microphone to the other end of the veranda in order to obtain any results at all. But by the time evening came we had successfully recorded not only the pygmy mongoose but Ticky as well, and had, in the bargain, unpacked ninety per cent of our equipment. So we bathed, changed, and dined, feeling well satisfied with ourselves.

After dinner we armed ourselves with a bottle of whisky and an abundant supply of cigarettes and, taking our pressure lamp, made our way to the Fon's house. The air was warm and drowsy, full of the scents of wood smoke and sun-baked earth. Crickets tinkled and trilled in the grass verges of the road, and in the gloomy fruit trees around the Fon's great courtyard we could hear the fruit bats honking and flapping their wings among the branches. In the courtyard a group of the Fon's children were standing in a circle clapping their hands and chanting in some sort of game, and away through the trees in the distance a small drum throbbed like an irregular heartbeat. We made our way through the maze of wives' huts, each lit by the red glow of a cooking fire, each redolent with the smell of roasting yams, frying plantain, stewing meat, or the sharp, pungent smell of dried salt fish. We came presently to the Fon's villa and he was waiting on the steps to greet us, looming large in the gloom, his robe swishing as he shook our hands.

"Welcome, welcome," he said, beaming. "Come, we go for inside."

"I done bring some whisky for make our heart happy," I said, flourishing the bottle as we entered the house.

"Wah! Good, good," said the Fon, chuckling. "Dis whisky na foine ting for make man happy."

He was wearing a wonderful scarlet and yellow robe that glowed like a tiger skin in the soft lamplight, and on one slender wrist was a thick, beautifully carved ivory bracelet. We sat down and waited in silence while the solemn ritual of the pouring of the first drink was observed. Then, when each of us was clutching a tumbler half full of neat whisky, the Fon turned to us, giving his wide, mischievous grin.

"Chirri-*ho!*" he said, raising his glass. "Tonight we go have happy time."

And so began what we were to refer to later as the Evening of the Hangover.

As the level in the whisky bottle fell, the Fon told us once again about his trip to Nigeria, how hot it had been and how much he had "shweated." His praise for the Queen knew no bounds for, as he pointed out, here was he in his own country feeling the heat and yet the Queen did twice the amount of work that he had done and always managed to look cool and charming. I found his lavish and perfectly genuine praise rather extraordinary, for the Fon belonged to a society where women are considered to be nothing more than rather useful beasts of burden.

"You like musica?" inquired the Fon of Jacquie, the subject of the Nigerian tour now being exhausted.

"Yes," said Jacquie, "I like it very much."

The Fon beamed at her. "You remember dis my musica?" he asked me.

"Yes, I remember. You get musica time no dere, my friend."

The Fon gave a prolonged crow of amusement. "You done write about dis my musica inside dis your book, eh?"

"Yes, that's right."

"And," said the Fon, coming to the point, "you done write about dis dancing an' dis happy time we done have, eh?"

"Yes. All dis dance we done do na fine one."

"You like we go show dis your wife what kind of dance we get here for Bafut?" he inquired, pointing a long forefinger at me.

"Yes, I like too much."

"Foine, foine . . . Come, we go for dancing house," he said, rising to his feet majestically, and stifling a belch with one slender hand. Two of his wives, who had been sitting quietly in the background, rushed forward and seized the tray of drinks and scuttled ahead of us as the Fon led us out of his house and across the compound towards his dancing house.

The dancing house was a great square building, not unlike the average village hall, but with an earth floor and very few and very small windows. At one end of the building was a line of wickerwork armchairs, which constituted a sort of royal enclosure, and on the wall above these were framed photographs of various members of the royal family. As we entered the dancing hall the assembled wives, about forty or fifty of them, uttered the usual greeting, a strange, shrill ululation, caused by yelling loudly and clapping their hands rapidly over their mouths at the same time. The noise was deafening. All the petty councillors there in their brilliant robes clapped their hands as well, and thus added to the general turmoil. Nearly deafened by this greeting, Jacquie and I were installed in two chairs, one on each side of the Fon. The table of drinks was placed in front of us, and the Fon, leaning back in his chair, surveyed us both with a wide and happy grin.

"Now we go have happy time," he said, and leaning forward poured out half a tumblerful of scotch each from the depths of a virgin bottle that had just been broached.

"Chirri-ho," said the Fon.

"Chin-chin," I said absent-mindedly.

"Na whatee dat?" inquired the Fon interestedly.

"What?" I asked, puzzled.

"Dis ting you say."

"Oh, you mean chin-chin?"

"Yes, yes, dis one."

"It's something you say when you drink."

"Na same same for chirri-ho?" asked the Fon, intrigued.

"Yes, na same same."

He sat silent for a moment, his lips moving, obviously comparing the respective merits of the two toasts. Then he raised his glass again.

"Shin-shin," said the Fon.

"Chirri-ho!" I responded, and the Fon lay back in his chair and went off into a paroxysm of mirth.

By now the band had arrived. It consisted of four youths and two of the Fon's wives, and the instruments consisted of three drums, two flutes, and a calabash filled with dried maize that gave off a pleasant rustling noise similar to a marimba. They got themselves organized in the corner of the dancing house, and then gave a few experimental rolls on the drums, watching the Fon expectantly. The Fon, having recovered from the joke, barked out an imperious order, and two of his wives placed a small table in the centre of the dance floor and put a pressure lamp on it. The drums gave another expectant roll.

"My friend," said the Fon, "you remember when you

done come for Bafut before you done teach me European dance, eh?"

"Yes," I said, "I remember."

It had been at one of the Fon's parties that, having partaken liberally of the Fon's hospitality, I had proceeded to show him, his councillors, and his wives how to do the conga. It had been a riotous success, but I had supposed that in the eight years that had passed the Fon would have forgotten about it.

"I go show you," said the Fon, his eyes gleaming. He barked out another order and about twenty of his wives shuffled out onto the dance floor and formed a circle round the table, each one holding firmly to the waist of the one in front. Then they assumed a strange crouching position, rather like runners at the start of a race, and waited.

"What are they going to do?" whispered Jacquie.

I watched them with an unholy glee. "I do believe," I said dreamily, "that he's been making them dance the conga ever since I left, and we're now going to have a demonstration."

The Fon lifted a large hand and the band launched itself with enthusiasm into a Bafut tune that had the unmistakable conga rhythm. The Fon's wives, still in their strange crouching position, proceeded to circle round the lamp, kicking their black legs out on the sixth beat, their brows furrowed with concentration. The effect was delightful.

"My friend," I said, touched by the demonstration, "dis na fine ting you do."

"Wonderful," agreed Jacquie enthusiastically. "They dance very fine."

"Dis na de dance you done teach me," explained the Fon.

"Yes, I remember."

He turned to Jacquie, chuckling. "Dis man your husband 'e get plenty power. We dance, we dance, we drink. . . . Wah! We done have happy time." The band came to an uneven halt, and the Fon's wives, smiling shyly at our clapping, raised themselves from their crouching position and returned to their former places along the wall. The Fon barked an order, and a large calabash of palm wine was brought in and distributed among the dancers, each getting her share poured into her cupped hands. Stimulated by this sight, the Fon filled all our glasses up again.

"Yes," he went on, reminiscently, "dis man your husband get plenty power for dance and drink."

"I no get power now," I said. "I be old man now."

"No, no, my friend," said the Fon laughing. "I be old, you be young."

"You look more young now den for the other time I done come to Bafut," I said, and really meant it.

"That's because you've got plenty wives," said Jacquie.

"Wah! No!" said the Fon, shocked. "Dis ma wives tire me too much."

He glared moodily at the array of females standing along the wall, and sipped his drink. "Dis ma wife dey humbug me too much," he went on.

"My husband says I humbug him," said Jacquie.

"Your husband catch lucky. 'E only get one wife, I get plenty," said the Fon, "an' dey de humbug me time no dere."

"But wives are very useful," said Jacquie.

The Fon regarded her sceptically.

"If you don't have wives you can't have babies. Men can't have babies," said Jacquie practically.

The Fon was so overcome with mirth at this remark I thought he might have a stroke. He lay back in his chair and laughed until he cried. Presently he sat up, wiping his eyes, still shaking with gusts of laughter. "Dis woman your wife get brain," he said, still chuckling, and poured Jacquie out an extra-large scotch to celebrate her intelligence. "You be good wife for me," he said, patting her on the head affectionately. "Shin-shin."

The band members now returned, wiping their mouths from some mysterious errand outside the dancing house and, apparently well fortified, launched themselves into one of my favourite Bafut tunes, the butterfly dance. It was a pleasant, lilting little tune, and the Fon's wives now took the floor and did the delightful dance that accompanied it. They danced in a row with minute but complicated hand and foot movements, and then the two that formed the head of the line joined hands, while the one at the farther end of the line whirled up and then fell backwards, to be caught and thrown upright again by the two with linked hands. As the dance progressed and the music got faster and faster the one representing the butterfly whirled more and more rapidly, and the ones with linked hands catapulted her upright again with more and more enthusiasm. Then, when the dance reached its feverish climax, the Fon rose majestically to his feet, amid screams of delight from the audience, and joined the end of the row of dancing wives. He started to whirl down the line, his scarlet and yellow robe turning into a blur of colour, loudly singing the words of the song.

"I dance, I dance, and no one can stop me," he carolled merrily, "but I must take care not to fall to the ground like the butterfly."

He went whirling down the line of wives like a top, his voice booming out above theirs.

"I hope to God they don't drop him," I said to Jacquie,

eyeing the two short, fat wives who, with linked hands, were waiting rather nervously at the head of the line to receive their lord and master.

The Fon performed one last mighty gyration and hurled himself backwards at his wives, who caught him neatly enough, but reeled under the shock. As the Fon landed he spread his arms wide so that for a moment his wives were invisible under the flowing sleeves of his robes and he lay there looking very like a gigantic multicoloured butterfly. He beamed at us, lolling across his wives' arms, his skull-cap slightly askew, and then his wives with an effort bounced him back to his feet again. Grinning and panting, he made his way back to us and hurled himself into his chair.

"My friend, na fine dance dis," I said in admiration, "you get power time no dere."

"Yes," agreed Jacquie, who had also been impressed by this display, "you get plenty power."

"Na good dance dis, na foine one," said the Fon, chuckling and automatically pouring us all out another drink.

"You get another dance here for Bafut I like too much," I said, "dis one where you dance with dat beer-beer for horse."

"Ah, yes, yes, I savvay um," said the Fon. "Dat one where we go dance with dis tail for horshe."

"That's right. Sometime, my friend, you go show dis dance for my wife?"

"Yes, yes, my friend," he said. He leaned forward and gave an order and a wife scuttled out of the dancing hall. The Fon turned and smiled at Jacquie. "Small time dey go bring dis tail for horshe an' den we go dance," he said.

Presently the wife returned, carrying a large bundle of white, silky horses' tails, each about two feet long, fixed in beautifully made handles woven out of leather thongs. The

Fon's tail was a particularly long and luxuriant one, and the thongs that had been used to make the handle were dyed blue, red, and gold. The Fon swished it experimentally through the air with languid, graceful movements of his wrist, and the hair rippled and floated like a cloud of smoke before him. Some twenty of the Fon's wives, each armed with their switches, went onto the floor and formed a circle. The Fon walked over and stood in the centre of the circle; he gave a wave of his horse's tail, the band struck up, and the dance was on.

Of all the Bafut dances this horsetail dance was undoubtedly the most sensuous and beautiful. The rhythm was peculiar, the small drums keeping up a sharp, staccato beat, while beneath them the big drums rumbled and muttered and the bamboo flutes squeaked and twittered with a tune that seemed to have nothing to do with the drums and yet merged with it perfectly. To this tune the Fon's wives gyrated slowly in a clockwise direction, their feet performing minute but formalized steps, while they waved the horses' tails gently to and fro across their faces. The Fon, meanwhile, danced round the inside of the circle in a counter-clockwise direction, bobbing, stamping, and twisting in a curiously stiff, unjointed sort of way, while his hand with incredibly supple wrist movements kept his horse's tail weaving through the air in a series of lovely and complicated movements. The effect was odd and almost indescribable; one minute the dancers resembled a bed of white seaweed, moved and rippled by sea movement, and the next minute the Fon would stamp and twist, stiff-legged, like some strange bird with white plumes, absorbed in a ritual dance of courtship among his circle of hens. Watching this slow pavan and the graceful movements of the tails had a curious sort of hypnotic effect, so that even when the dance ended with a roll of drums you could still see the white tails weaving and merging before your eyes.

The Fon moved gracefully across the floor towards us, twirling his horse's tail negligently, and sank into his seat. He beamed breathlessly at Jacquie. "You like dis ma dance?" he asked.

"It was *beautiful*," she said. "I liked it very much."

"Good, good," said the Fon, well pleased. He leaned forward and inspected the whisky bottle hopefully, but it was obviously empty. Tactfully I refrained from mentioning that I had some more over at the rest house. The Fon surveyed the bottle gloomily. "Whisky done finish," he pointed out.

"Yes," I said unhelpfully.

"Well," said the Fon, undaunted, "we go drink gin."

My heart sank, for I had hoped that we could now move on to something innocuous like beer to quell the effects of so much neat alchohol. The Fon roared at one of his wives and she ran off and soon reappeared with a bottle of gin and one of bitters. The Fon's idea of gin-drinking was to pour about half a tumblerful and then colour it a deep brown with bitters. The result was guaranteed to slay an elephant at twenty paces. Jacquie, on seeing this cocktail the Fon concocted for me, hastily begged to be excused, saying that she couldn't drink gin on doctor's orders. The Fon, though obviously having the lowest possible opinion of a medical man who could even suggest such a thing, accepted with good grace.

The band started up again and everyone poured onto the floor and started to dance, singly and in couples. As the rhythm of the tune allowed it, Jacquie and I got up and did a swift foxtrot round the floor, the Fon roaring encouragement and his wives hooting with pleasure.

"Foine, foine," shouted the Fon as we swept past.

"Thank you, my friend," I shouted back, steering Jacquie carefully through what looked like a flower bed of councillors in their multicoloured robes.

"I do wish you wouldn't tread on my feet," said Jacquie plaintively.

"Sorry. My compass bearings are never at their best at this hour of night."

"So I notice," said Jacquie acidly.

"Why don't you dance with the Fon?" I inquired.

"I did think of it, but I wasn't sure whether it was the right thing for a mere woman to ask him."

"I think he'd be tickled pink. Ask him for the next dance," I suggested.

"What can we dance?" asked Jacquie.

"Teach him something he can add to his Latin American repertoire," I said. "How about a rumba?"

"I think a samba would be easier to learn at this hour of night," said Jacquie. So when the dance ended we made our way back to where the Fon was sitting, topping up my glass.

"My friend," I said, "you remember dis European dance I done teach you when I done come for Bafut before?"

"Yes, yes, na foine one," he replied, beaming.

"Well, my wife like to dance with you and teach you other European dance. You agree?"

"Wah!" bellowed the Fon in delight. "Foine, foine. Dis your wife go teach me. Foine, foine, I agree."

Eventually we discovered a tune that the band could play that had a vague samba rhythm, and Jacquie and the Fon rose to their feet, watched breathlessly by everyone in the room.

The contrast between the Fon's six foot three and Jacquie's five foot one made me choke over my drink as they

took the floor. Very rapidly Jacquie showed him the simple, basic steps of the samba, and to my surprise the Fon mastered them without trouble. Then he seized Jacquie in his arms and they were off. The delightful thing from my point of view was that as he clasped Jacquie tightly to his bosom she was almost completely hidden by his flowing robes, so at some points in the dance she could not be seen at all and it looked as though the Fon, having mysteriously grown another pair of feet, were dancing round by himself. There was something else about the dance that struck me as curious, but I could not think what it was for some time. Then I suddenly realized that Jacquie was leading the Fon. They sambaed past, both grinning at me, obviously enjoying themselves.

"You dance fine, my friend," I shouted. "My wife done teach you fine."

"Yes, yes," roared the Fon over the top of Jacquie's head. "Na foine dance dis. Your wife na good wife for me."

Eventually, after half an hour's dancing, they returned to their chairs, hot and exhausted. The Fon took a large gulp of neat gin to restore himself, and then leaned across to me.

"Dis your wife na foine," he said in a hoarse whisper, presumably thinking that praise might turn Jacquie's head. "She dance foine. She done teach me foine. I go give her mimbo. Special mimbo I go give her."

I turned to Jacquie who, unaware of her fate, was sitting fanning herself. "You've certainly made a hit with our host," I said.

"He's a dear old boy," said Jacquie, "and he dances awfully well. Did you see how he picked up that samba in next to no time?"

"Yes," I said, "and he was so delighted with your teaching that he's going to reward you."

Jacquie looked at me suspiciously. "How's he going to reward me?" she asked.

"You're now going to receive a calabash of special mimbo—palm wine."

"Oh, God, and I can't stand the stuff," said Jacquie in horror.

"Never mind. Take a glassful, taste it, tell him it's the finest you've ever had, and then ask if he will allow you to share it with his wives."

Five calabashes were brought, each with the neck plugged with green leaves, and the Fon solemnly tasted them all before making up his mind which was the best vintage. Then a glass was filled and passed to Jacquie. Summoning up all her social graces, she took a mouthful, rolled it round her mouth, swallowed, and allowed a look of intense satisfaction to appear on her face. "This is very fine mimbo," she proclaimed in delighted astonishment, with the air of one who has just been presented with a glass of Napoleon brandy. The Fon beamed. Jacquie took another sip, as he watched her closely. An even more delighted expression appeared on her face. "This is the best mimbo I've ever tasted," said Jacquie.

"Ha! Good!" said the Fon. "Dis na foine mimbo. Na fresh one."

"Will you let your wives drink with me?" asked Jacquie.

"Yes, yes," said the Fon with a lordly wave of his hand, and so the wives shuffled forward, grinning shyly, and Jacquie hastily poured the remains of the mimbo into their pink palms.

At this point, the level of the gin bottle having fallen alarmingly, I suddenly glanced at my watch and saw, with horror, that in two and a half hours it would be dawn. So, pleading heavy work on the morrow, I broke up the party.

The Fon insisted on accompanying us to the foot of the steps that led up to the rest house, preceded by the band. Here he embraced us fondly.

"Good night, my friend," he said, shaking my hand.

"Good night," I replied. "Thank you. You done give us happy time."

"Yes," said Jacquie. "Thank you very much."

"Wah!" said the Fon, patting her on the head. "We done dance foine. You be good wife for me, eh?"

We watched him as he wended his way across the great courtyard, tall and graceful in his robes, the boy trotting beside him carrying the lamp that cast a pool of golden light about him. They disappeared into the tangle of huts, and the twittering of the flutes and the bang of the drums became fainter and died away, until all we could hear was the calls of crickets and tree frogs and the faint honking cries of the fruit bats. Somewhere in the distance the first cock crowed, huskily and sleepily, as we crept under our mosquito nets.

MAIL BY HAND

My good friend,
  Good morning to you all.
  Your note to me received and contents well understood.
  I am a beat relief from that cough but not much.
  I agree for you to hire my Landrover as from today on
weekly payments. I will also want to bring to your notice
that the Landrover will be under your charge as from to-
day, but any time I am called for a meeting at Ndop,
Bemenda or elsewhere, or any urgent matter, I shall in-
form you to allow me the motor for the day.
  I want to remind you of the last trip which you hired the
Landrover and settlement had not yet been made.
                                    Your good friend,
                                    Fon of Bafut

4: BEEF IN BOXES

As soon as Bob and Sophie had joined us in Bafut we set about the task of organizing our already large and ever-growing collection. The great, shady veranda that ran round the upstairs rooms of the Fon's rest house was divided into three sections, one for reptiles, one for birds, and one for mammals. Thus each of us had a particular section to look after, and whoever finished first lent a hand with somebody else's group. First thing in the morning we would all wander to and fro along the veranda in our pyjamas, looking at each animal to make sure it was all right. It is only by this day-to-day routine of careful watching that you can get to know your animals so well that the slightest sign of illness makes itself apparent to you, when to anyone else the animal would appear to be perfectly healthy and normal. Then we would clean and feed all the delicate animals that could not wait (such as the sunbirds, which had to have their nectar as soon as it was light, and the baby creatures that needed their early-morning bottles), and then we would pause for breakfast.

It was during mealtime that we would compare notes on our charges. This mealtime conversation would have put any normal mortal off his feed, for it was concerned mainly with the bowel movements of our creatures; with wild animals, diarrhoea or constipation can be a good indication as to whether or not you are feeding it correctly, and also it can be the first (and sometimes the only) symptom of an illness.

On any collecting trip, acquiring the animals is, as a rule, the simplest part of the job. As soon as the local people discover that you are willing to buy live wild creatures the stuff comes pouring in; ninety per cent is, of course, the commoner species, but there is an occasional rarity. If you want the really rare stuff you generally have to go out and find it yourself, but while you are devoting your time to this you can be sure that all the common local fauna will be brought in to you. So, one might almost say that getting the animals is easy; the really difficult part is keeping them.

The chief thing you have to contend with when you have got a newly caught animal is not so much that it might be suffering from shock at being captured, but that the capture has resulted in its having to exist in close proximity to a creature it regards as an enemy of the worst possible sort: yourself. On many occasions you will get an animal that takes to captivity beautifully, but can never reconcile itself to the fact that it has to exist on such intimate terms with man. This is the first great barrier that you have to break down, and you can do it only by patience and kindness. For month after month an animal may snap and snarl at you every time you approach its cage, until you begin to despair of ever making any favourable impression on it. Then, one day, sometimes without any preliminary warning, it will trot forward and take food from your hand, or allow you to tickle it behind the ears. At such moments you feel that all the waiting has been justified.

Feeding, of course, is one of your main problems. Not only do you have to have a fairly extensive knowledge of what each species you are liable to encounter eats in the wild state, but you have to work out a suitable substitute if the natural food is unavailable, and then teach your specimen to eat it. Also, you have to cater for their individual likes and dislikes, which vary enormously. I have known a rodent which, refusing all normal rodent food—such as fruit, bread, vegetables—lived for three days on an exclusive diet of spaghetti. I have had a group of five monkeys, of the same age and species, who displayed the most weird idiosyncrasies. Two of the five had a passion for hard-boiled eggs, while the other three were frightened of the strange white shapes and refused to touch them, actuaally screaming in fear if you introduced such a fearsome object as a hard-boiled egg into their cage. These five monkeys all adored oranges but, whereas four would carefully peel their fruit and throw away the skin, the fifth would peel his orange equally carefully and then throw away the orange and eat the peel. When you have a collection of several hundred creatures, all displaying such curious characteristics, you are sometimes nearly driven mad in your efforts to satisfy their desires and so keep them healthy and happy.

But of all the irritating and frustrating tasks that you have to undertake during a collecting trip, the hand-rearing of baby animals is undoubtedly the worst. To begin with, they are generally incredibly stupid over taking a bottle, and there is nothing quite so unattractive as to struggle with a baby animal in a sea of lukewarm milk. Second, they have to be kept warm, especially at night, and this means (unless you take them to bed with you, which is generally the answer) that you have to get up several times during the night to replenish hot-water bottles. After a hard day's work, to have to drag yourself out of bed at three in the morning to fill hot-water bottles is an occupation that soon loses its charm. Third, all baby animals have extremely delicate stomachs, so you have to watch them like a hawk to make sure that the milk you are giving them is not too rich or too weak, for with the former they can develop intestinal troubles which may lead to nephritis, which will probably kill them, and the latter can lead to loss of weight and condition, which leaves the animal open to all sorts of fatal complaints.

Contrary to my gloomy prognostications, the baby black-eared squirrel, Squill-lill Small (Small to her friends), proved an exemplary baby. During the day she lay twitching in a bed of cotton wool balanced on a hot-water bottle in the bottom of a deep biscuit tin, and at night the tin was placed by our beds under the rays of a Tilley infra-red heater. Almost immediately we were made aware of the fact that Small had a will of her own. For such a tiny animal she could produce an extraordinary volume of noise, her cry being a loud and rapid series of "chucks" that sounded like a cheap alarm clock going off. Within the first twenty-four hours she had learned at what times to expect her feedings, and if we were so much as five minutes late she would trill and chuck incessantly until we arrived with the food. Then came the great day when Small's eyes opened for the first time and she could take a look at her foster parents and the world in general. This, however, presented a new problem. The day her eyes opened we happened to be a bit late with her food. We had rather dawdled over our own lunch, deep in a discussion about some problem or other, and we had, I regret to say, forgotten all about Small. Suddenly I heard a faint scuffling behind me, turned round, and saw Small squatting in the doorway of the room we used as a dining room, looking, to say the least, extremely indignant. As soon as she saw us she went off like an alarm clock and, hurrying across the floor, hauled herself, panting, up the chair Jacquie was sitting on, and then leaped to her shoulder, where she sat flicking her tail up and down and shouting indignantly into Jacquie's ear. Now this, for a baby squirrel, was quite a feat. To begin with, as I say, her eyes had only just opened. Yet she had succeeded in hauling herself out of her tin, found her way out of our bedroom (piled high with camera equipment and film), made her way down the full length of the veranda, running the gauntlet of any number of cages filled with potentially dangerous beasts, and eventually located us (presumably by sound) in the dining room which was at the extreme end of the veranda. So she had covered some seventy yards over unknown territory, through innumerable dangers, in order to tell us she was hungry. Needless to say, she got the praise which was her due and, what was more important from her point of view, she got her lunch.

As soon as Small's eyes opened she began to grow rapidly, and she soon developed into one of the loveliest squirrels I have ever seen. Her orange head and neat, black-rimmed ears nicely set off her large dark eyes, and her fat body developed a rich moss-green tinge against which the two lines of white spots that decorated her sides stood out like a cat's eyes on a dark road. But it was her tail that was her best feature, I think. Long and thick, green above and vivid orange below, it was a beautiful sight. She loved to sit with it curved over her back, the tip actually hanging over her nose, and then she would flick it gently with an undulating movement so that the whole thing looked like a candle flame in a draught.

Even when she was quite grown up, Small slept in her biscuit tin by our bed. She would awake early in the morning and, uttering her loud cry, would leap from the tin onto one of our beds and crawl under the bedclothes with us. Having spent ten minutes or so investigating our semi-comatose bodies, she would jump to the floor and go on an expedition on the veranda. From these expeditions she would frequently return with some treasure she had found (such as a bit of rotten banana, or a dry leaf, or a bougainvillaea flower) and store it somewhere in our beds, getting most indignant if we hurled the offering out onto the floor. This went on for some months, and then came the day when I decided that Small would have to occupy a cage like the rest of the animals, for I awoke one morning in excruciating agony to find her doing her best to stuff a peanut into my ear. Having found such a delicacy on the veranda, she obviously thought that to simply cache it in my bed was not safe enough, and there was my ear providing an ideal hiding place.

Bug-Eyes, the needle-clawed lemur we had captured near Eshobi, was another baby, although she was fully weaned when we got her. She had become tame in a very short space of time and very rapidly became one of our favourites. For her size she had enormous hands and feet, with long, attenuated fingers, and to see her dancing around her cage on her hind legs, her immense hands held up as though in horror, her eyes almost popping out of her head, as she pursued a moth or butterfly we had introduced, was a delightfully comic sight. Once she had caught it she would sit there with it clasped tightly in her pink hand and regard it with a wild, wide-eyed stare, as if amazed that such a creature should suddenly appear in the palm of her hand. Then she would stuff it into her mouth and sit there with what appeared to be a fluttering mous-

tache of butterfly wing decorating her face, over which her huge eyes peered in astonishment.

It was Bug-Eyes who first showed me an extraordinary habit that bushbabies have, a habit which, to my shame, I had never noticed before, in spite of the fact that I had kept innumerable bushbabies. I was watching her one morning when she had popped out of her nesting box for a feed of mealworms and a quick wash and brush-up. She had, as I said before, large ears which were as delicate as flower petals. They were so fine that they were semi-transparent and, presumably to prevent them from becoming torn or damaged in the wild state, she had the ability to fold them back against the sides of her head like the furled sails of a yacht. Now her ears were terribly important to her, as you could tell by watching her. The slightest sound, however faint, would be picked up and her ears would twitch and turn towards it like radar. Now, I had always noticed that she spent a lot of time cleaning and rubbing her ears with her hands, but on this particular morning I watched the whole process from start to finish and was considerably startled by what I saw. She sat on a branch, staring dreamily into space, and daintily cleaned her tail, parting the hair carefully and making sure there were no snags or tangles, reminding me of a little girl plaiting her hair. Then she put one of her outsize, puppet-like hands beneath her and deposited in the palm a drop of urine. With an air of concentration, she rubbed her hands together and then proceeded to anoint her ears with the urine, rather after the manner of a man rubbing brilliantine into his hair. Then she got another drop of urine and rubbed it carefully over the soles of her feet and the palms of her hands, while I sat and watched her in amazement.

I watched her do this three days in succession before I was satisfied that I was not imagining things, for it seemed to me to be one of the weirdest animal habits I had ever encountered. I can conclude only that the reason for it was this: The skin of the ears was, as I say, extremely delicate and thin. Unless it was kept moistened it must inevitably get dry and perhaps crack, which would have been fatal for an animal that relied so much on hearing. The same would apply to the delicate skin on the soles of her feet and hands, but here the urine would also provide an additional advantage. The soles of feet and hands were slightly cupped, so that as the creature leaped from bough to bough the hands and feet acted almost like the suckers on the toes of a tree frog. Now, moistened with urine, these "suckers" became twice as efficient. When, later on in the trip, we obtained a great number of Demidoff's bushbabies (the smallest of the tribe, each being the size of a large mouse) I noticed that they all had the same habit.

This is, to my mind, the best part of a collecting trip, the close daily contact with the animals that allows you to observe, learn, and record. Every day, and almost at every moment of the day, something new and interesting was happening somewhere in the collection. The following diary entries show fairly well how each day bristled with new tasks and interesting observations.

14/2/57: Two patas monkeys brought in; both had severe infestation of jiggers in toes and fingers. Had to lance them, extract jiggers, and as precaution against infection injected penicillin. Baby civet did her first adult "display," making the mane of hair on her back stand up when I approached her cage suddenly. She accompanied this action with several loud sniffs, much deeper and more penetrating than her normal sniffing around food. Large brow-leaf toad brought in with extraordinary eye trouble. What appears to be a large malignant growth, situated

behind the eyeball, had blinded the creature and then
grown outwards, so that the toad looked as though it
were wearing a large balloon over one eye. It did not
appear to be suffering so am not attempting to remove
the growth. So much for animals being happy and care-
free in the wild state.

20/2/57: At last, after much trial and error, Bob has discovered what the hairy frogs eat: snails. We had previously
tried young mice and rats, baby birds, eggs, beetles and
their larvae, locusts, all without success. Snails they devour
avidly, so we have high hopes of getting the frogs back
alive. Have had an outbreak of what appears to be
nephritis among the Demidoff's bushbabies. Two discovered this morning drenched in urine as though they
had been dipped. Have weakened the milk they get; it
may be too strong. Also organized more insect food for
them. The five baby Demidoff's are still thriving on their
Complan milk, which is curious, as this is incredibly rich,

and if ordinary dried milk affects the adults, one would have thought Complan would have had a similar effect on the babies.

16/3/57: Two nice cobras brought in, one about six feet long and the other about two feet. Both fed straight away. Best item today was female pygmy mongoose and two babies. The babies are still blind and an extraordinary pale fawn in contrast to the dark brown mother. Have removed babies to hand-rear them, as felt sure female would either neglect or kill them if they were put with her.

17/3/57: Young pygmy mongooses flatly refuse to feed from bottle or from fountain-pen filler. In view of this (since their chances seemed slim) put them into cage with female. To my surprise she has accepted them and is suckling them well. Most unusual. Had two broken leg jobs today: Woodford's owl, which had been caught in a gin trap, and a young hawk with a greenstick fracture. I don't think the owl will regain use of leg, for all the ligaments appear to be torn, and the bone badly splintered. Hawk's leg should be O.K. as it's a young bird. Both are feeding well. Demidoff's make a faint mewing hiss when disturbed at night, the only sound I have heard them make, apart from their batlike twittering when fighting. Clawed toads have started to call at night: very faint "peep-peep" noise, rather like someone flicking the edge of a glass gently with a finger-nail.

2/4/57: Young male chimp, about two years old, brought in today. Was in a terrible mess. Had been caught in one of the wire noose traps they use for antelopes, and had damaged its left hand and arm. The palm of the hand and the wrist were split right open and badly infected with gangrene. The animal was very weak, not being able to sit up, and the colour of the skin was a curious yellowish grey. Attended to wound and injected penicillin. Drove it in to Bemenda for the Dept. of Agriculture's vet

to have a look at, as did not like skin colour or curious
lethargy in spite of stimulants. He took blood test and
diagnosed sleeping sickness. Have done all we can but the
animal appears to be sinking fast. He seems pathetically
grateful for anything you do for him.

3/4/57: Chimp died. This is a "protected" animal, and yet
up here, as in other parts of the Cameroons, they are
killed and eaten regularly. Big rhinocerus viper fed for
first time: small rat. One of the Green Forest squirrels
appears to be developing a bald patch on his back: pre-
sume lack of vitamins so am increasing his Abidec. As we
now get good supply of weaver bird's eggs each day, all
the squirrels are getting them, in addition to their normal
diet. The brush-tail porcupines, when disturbed at night,
beat rapid tattoo with their hind feet (like a wild rabbit)
then swing their backside round to face danger and
rustle bunch of quills on end of tail, producing a sound
reminiscent of rattlesnake.

5/4/57: Have found simple rapid way of sexing pottos.
Nice young male brought in today. Although external
genitalia in both sexes is remarkably similar at a super-
ficial glance, have discovered that simplest way is to smell
them. The testicles of the male give out a faint, sweet
odour, like pear drops, when the animal is handled.

We were not the only ones who were interested in the
animals. Many of the local people had never seen some of
the creatures we had acquired, and many called and asked
for permission to look round the collection. Then, one day,
the headmaster of the local mission school called and
asked if he could bring his entire school of two hundred-
odd boys to see the collection. I was glad to agree to this,
for I feel that if you can, by showing live animals, arouse
people's interest in their local fauna and its preservation
you are doing something worth while. So, on the appointed

date, the boys came marching down the road in a double
column, shepherded by five masters. In the road below the
rest house the boys were divided up into groups of twenty
and then brought up by a master. Jacquie, Sophie, Bob,
and I took up stations at various points in the collection to
answer any queries. The boys behaved in a model fashion;
there was no pushing or shoving, no skylarking. They
wended their way from cage to cage, absorbed and fascinated, uttering amazed cries of "Wah!" at each new wonder and clicking their fingers in delight. Finally, when the
last group had been led round, the headmaster grouped all
the boys at the bottom of the steps and then turned to me,
beaming.

"Sir," he said, "we are very grateful to you for allowing
us to see your zoological collections. May I ask if you would
be kind enough to answer some of the boys' questions?"

"Yes, with pleasure," I said, taking up my stand on the
steps above the crowd.

"Boys," roared the Head, "Mr. Durrell has kindly say
that he will answer any questions. Now who has a question?"

The sea of black faces below me screwed themselves up
in thought, tongues protruded, toes wiggled in the dust.
Then, slowly at first, but with increasing speed as they lost
their embarrassment, they shot questions at me, all of
which were extremely intelligent and sensible. There was,
I noticed, one small boy in the front of the crowd who had,
throughout the proceedings, fixed me with a basilisk eye.
His brow was furrowed with concentration, and he stood
stiffly at attention. At last, when the supply of questions
started to peter out, he suddenly summoned up all his cour-
age and shot his hand up.

"Yes, Uano, what is your question?" asked the Head,
smiling down fondly at the boy.

The boy took a deep breath and then fired his question at me rapidly. "Please, sah, can Mr. Durrell tell us why he take so many photographs of the Fon's wives?"

The smile vanished from the Head's face and he threw me a look of chagrin. "That is not a zoological question, Uano," he pointed out severely.

"But please sah, why?" repeated the child stubbornly.

The Head scowled ferociously. "That is *not* a zoological question," he thundered. "Mr. Durrell only said he would answer zoological questions. The matter of the Fon's wives is not zoological."

"Well, loosely speaking it could be called biological, Headmaster, couldn't it?" I asked, coming to the lad's rescue.

"But, sir, they shouldn't ask you questions like that," said the Head, mopping his face.

"Well, I don't mind answering. The reason is that, in my country, everyone is very interested to know how people in other parts of the world live and what they look like. I can tell them, of course, but it's not the same as if they see a photograph. With a photograph they know exactly what everything is like."

"There," said the Headmaster, running a finger round the inside of his collar. "There, Mr. Durrell has answered your question. Now, he is a very busy man so there is no more time for further questions. Kindly get into line."

The boys formed themselves once more into two orderly lines, while the Headmaster shook my hand and earnestly assured me that they were all most grateful. Then he turned once more to the boys.

"Now, to show our appreciation to Mr. Durrell I want three hearty cheers."

Two hundred young lungs boomed out the hearty

cheers. Then the boys at the head of the line produced from bags they were carrying several bamboo flutes and two small drums. The Headmaster waved his hand and they started to walk off down the road, led by the school band playing, of all things, "Men of Harlech." The Head followed them, mopping his face, and from the dark looks he kept darting at young Uano's back it did not argue well for the boy's prospects when he got back to the classroom.

That evening the Fon came over for a drink and, after we had shown him the new additions to the collection, we sat on the veranda and I told him about Uano's zoological question.

The Fon laughed and laughed, particularly at the embarrassment of the Headmaster. "Why you never tell um," he inquired, wiping his eyes, "why you never tell um dat you take dis photo of dis ma wife for show all Europeans for your country dat Bafut women be beautiful?"

"Dis boy na picken," I said solemnly. "I think sometime he be too small to understand dis woman palava."

"Na true, na true," said the Fon, chuckling, " 'e be picken. 'E catch lucky, 'e no get women for humbug him."

"They tell me, my friend," I said, trying to steer the conversation away from the pros and cons of married life, "they tell me tomorrow you go for N'dop. Na so?"

"Na so," said the Fon. "I go for two days, for Court. I go come back for morning time tomorrow tomorrow."

"Well," I said, raising my glass, "safe journey, my friend."

The following morning, clad in splendid yellow and black robes and wearing a curious hat, heavily embroidered, with long, drooping earflaps, the Fon took his seat in the front of his new Land-Rover. Into the back went the necessities of travel: three bottles of scotch, his favourite wife, and three council members. He waved vigorously to us until the vehicle rounded the corner and was lost from sight.

That evening, having finished the last chores of the day, I went out onto the front veranda for a breath of air. In the great courtyard below I noticed large numbers of the Fon's children assembling. Curiously I watched them. They grouped themselves in a huge circle in the centre of the compound and, after much discussion and argument, started to sing and clap their hands rhythmically, accompanied by a seven-year-old who stood in the centre of the circle beating a drum. Standing like this, they lifted up their young voices and sang some of the most beautiful and haunting of the Bafut songs. This, I could tell, was not just an ordinary gathering of children; they had assembled there for some definite purpose, but what they were celebrating (unless it was their father's departure) I could not think. I stood there watching them for a long time and then John, our houseboy, appeared at my elbow in the unnervingly silent way he had.

"Dinner ready, sah," he said.

"Thank you, John. Tell me, why all dis picken sing for the Fon's compound?"

John smiled shyly. "Because de Fon done go for N'dop, sah."

"Yes, but why they sing?"

"If the Fon no be here, sah, each night dis picken must for sing inside de Fon's compound. So dey keep dis his compound warm."

This, I thought, was a delightful idea. I peered down at the circle of children, singing lustily in the gloomy wastes of the great courtyard, to keep their father's compound warm.

"Why they never dance?" I asked.

"Dey never get light, sah."

"Take them the pressure light from the bedroom. Tell them I send it so that I can help keep the Fon's compound warm."

"Yes, sah," said John, grinning delightedly.

He hurried off to fetch the light and presently I saw it cast a golden pool round the circle of children. There was a pause in the singing, while John delivered my message, and then came a series of delighted shrieks and echoing up to me the shrill voices crying, "Tank you, Masa, tank you."

As we sat down to dinner the children were singing like larks, and stamping and weaving their way round the lamp, their shadows long and attenuated, thrown halfway across the courtyard by the softly hissing lamp in their midst.

MAIL BY HAND

My good friend,
    Would you like to come and have a drink with us this
evening at eight o'clock?
                                          Your friend,
                                          Gerald Durrell

My good friend,
    Expect me a 7.30 p.m. Thanks.
                                          Your good friend,
                                          Fon of Bafut

5: FILM-STAR BEEF

There are several different ways of making an animal film, and probably one of the best methods is to have a team of cameramen who spend about two years in some tropical part of the world filming the animals in their natural state. Unfortunately this method is expensive, and unless you have the time and the resources of Hollywood behind you it is impossible to employ.

For someone like myself, with only a limited amount of time and money to spend in a country, the only way to film animals is under controlled conditions. The difficulties of trying to film wild animals in a tropical forest are enough to make even the most ardent photographer grow pale. To begin with, you hardly ever see a wild animal and, when you do, you generally have only a momentary glimpse as it scuttles off into the undergrowth. To be in the right spot at the right time with your camera set up, your exposure correct, and an animal in front of you in a suitable setting, indulging in some interesting and filmable action, would be almost a miracle. So, the only way round this is to catch your animal first and establish it in captivity. Once it has lost some of its fear of human beings you can begin work. Inside a huge netting "room" you create a scene which is as much like the animal's natural habitat as possible, and yet is—photographically speaking—suitable. That is to say, it must not have too many holes in which a shy creature can hide, your undergrowth must not be so thick that you get awkward patches of shade, and so on. Then you introduce your animal to the set, and allow it time to settle down. This may take anything from an hour to a couple of days.

It is essential, of course, to have a good knowledge of the animal's habits, and to know how it will react under certain circumstances. For example, a hungry pouched rat, if released in an appropriate setting and finding a lavish selection of forest fruits on the ground, will promptly proceed to stuff as many of them into his immense cheek pouches as they will hold, so that in the end he looks as though he's suffering from a particularly virulent attack of mumps. So, if you don't want to end up with nothing more exciting than a series of pictures of some creature wandering aimlessly to and fro amid bushes and grass, you must provide the circumstances which will allow it to display some interesting habit or action. However, even when you have reached this stage you still require two other things: patience and luck. An animal—even a tame one—cannot be told what to do like a human actor. Sometimes a creature which has performed a certain action day after day for weeks will, when faced by the camera, develop an acute attack of stage fright, and refuse to perform. When you have spent hours in the hot sun getting everything ready, to be treated to this sort of display of temperament makes one feel positively homicidal.

A prize example of the difficulties of animal photography was, I think, the day we attempted to photograph the water chevrotain. This delightful little antelope is about the size of a fox terrier, with a rich chestnut-coloured coat handsomely marked with streaks and spots of white. Small, dainty and beautifully marked, the water chevrotain is extremely photogenic. There are several interesting things about the chevrotain, one of which is its adaptation to a semi-aquatic life in the wild state. Chevrotains spend most of their time wading and swimming in streams in the forest, and can even swim for considerable distances under water. The second curious thing is that they have a passion for snails and beetles, and such carnivorous habits in an antelope are, to say the least, unusual. The third thing about them is their extraordinary placidity and tameness: I have known a chevrotain, an hour after capture, to take food from my hand and allow me to tickle its ears, for all the world as if it had been born in captivity.

Our water chevrotain was no exception; she was ridiculously tame, adored having her head and tummy scratched, and would devour, with every evidence of satisfaction, any quantity of snails and beetles you cared to provide. Apart from this she spent her spare time trying to bathe in her water bowl, into which she could just jam—with considerable effort—the extreme rear end of her body.

So, in order to show off her carnivorous and aquatic habits, I designed a set embracing a section of river bank. The undergrowth was carefully placed so that it would show off her perfect adaptive colouration to the best advantage. One morning, when the sky was free from cloud and the sun was in the right direction, we carried the

chevrotain cage out to the set and prepared to release her.

"The only thing I'm afraid of," I said to Jacquie, "is that I'm not going to get sufficient movement out of her. You know how quiet she is. She'll probably walk into the middle of the set and refuse to move."

"Well, if we offer her a snail or something from the other side I should think she'll walk across," said Jacquie.

"As long as she doesn't just stand there, like a cow in a field. I want to get *some* movement out of her."

I got considerably more movement out of her than I anticipated. The moment the slide of her cage was lifted she stepped out daintily and paused with one slender hoof raised. I started the camera and awaited her next move. Her next move was, to say the least, unexpected. She shot across my carefully prepared set like a rocket, went right through the netting wall as if it had not been there, and disappeared into the undergrowth in the middle distance before any of us could make a move to stop her. Our reactions were slow, because this was the last thing we had expected, but as I saw my precious chevrotain disappearing from view I uttered such a wail of anguish that everyone, including Phillip the cook, dropped whatever they were doing and assembled on the scene like magic.

"Water beef done run," I yelled. "I go give ten shillings to the man who go catch um."

The effect of this lavish offer was immediate. A wedge of Africans descended onto the patch of undergrowth into which the antelope had disappeared, like a swarm of hungry locusts. Within five minutes Phillip, uttering a sergeant-major-like roar of triumph, emerged from the bushes clutching the kicking, struggling antelope to his bosom. When we replaced her in her cage she stood quite quietly, gazing at us with limpid eyes as if astonished at all the fuss. She licked my hand in a friendly fashion, and when tickled behind the ears went off into her usual trancelike state, with half-closed eyes. We spent the rest of the day trying to film the wretched creature. She behaved beautifully in her box, splashing in a bowl of water to show how aquatic she was, eating beetles and snails to show how carnivous she was, but the moment she was released into the film set she fled towards the horizon as if she had a brace of leopards on her trail. At the end of the day, hot and exhausted, I had exposed fifty feet of film, all of which showed her standing stock still outside her box, preparatory to dashing away. Sadly we carried her cage back to the rest house, while she lay placidly on her banana-leaf bed and munched beetles. It was the last time we tried to photograph the water chevrotain.

Another creature that caused me untold anguish in the

photographic field was a young Woodford's owl called, with singular lack of originality, Woody. Woodfords are very lovely owls, with a rich chocolate plumage splashed and blotched with white, and possessed of what must be the most beautiful eyes in the whole of the owl family. They are large, dark, and liquid, with heavy lids that are a delicate pinkish mauve. These they raise and lower over their eyes in what seems to be slow motion, like an ancient film actress cogitating whether to make a comeback. This seductive eyelid fluttering is accompanied by loud, castanet-like clickings of the beak. When the owls are excited the eyelid fluttering becomes very pronounced and they sway from side to side on the perch, as if about to start a hula, and then they will suddenly spread their wings and stand there clicking their beaks at you, looking like a tombstone angel of the more fierce and religious variety. Now Woody would perform all these actions perfectly inside his cage and would, morover, perform them to order when shown a succulent titbit such as a small mouse. I felt sure that, if he were provided with a suitable background, I could get his display on film with the minimum of trouble.

So in the netting room I used for bird photography I set to and created what looked like a forest tree, heavily overgrown with creepers and other parasites, and with green leaves and a blue sky as background. Then I carried Woody out and placed him on the branch in the midst of this wealth of foliage. Now, the action I wanted him to perform was a simple and natural one which should not, I would have thought, tax even the brain of an owl. With a little cooperation on his part the whole thing could have been over in ten minutes. He sat on the branch regarding us with wide-eyed horror, while I took up my position behind the camera. Just as I pressed the button he blinked

his eyes once, very rapidly, and then, as if overcome with disgust at our appearance, he very firmly turned his back on us and the camera. Trying to remember that patience was the first requisite of an animal photographer, I wiped the sweat from my eyes, walked up to the branch, turned him round, and walked back to the camera. By the time I had reached it Woody once more had his back towards us. I thought that maybe the light was too strong, so several members of the staff were sent to cut branches, and these were rigged up so that the bird was sheltered from the direct rays of the sun. But still he persisted in keeping his back turned towards us. It was obvious that if I wanted to photograph him I would have to rearrange my set so that it faced the opposite way. With considerable labour about a ton of undergrowth was carefully shifted and re-arranged, so that Woody was now facing the way he obviously preferred. During this labour, while we sweated with massive branches and coils of creepers, he sat there, regarding us in wide-eyed surprise. He generously allowed me to get the camera set up in the right position (a complicated job, for I was now shooting almost directly into the sun) and then he calmly turned his back on it. I could have strangled him. By this time ominous black clouds were rolling up, preparatory to obscuring the sun, and so further attempts at photography were impossible. I packed up the camera and walked to the branch, murder in my heart, to collect my star. As I approached he turned round, clicked his beak delightedly, executed a rapid hula, and then spread his wings and bowed to me, with the mock-shy air of an actor taking his seventeenth curtain call.

Of course, not all our stars caused us trouble. In fact, one of the best sequences I managed to get on film was accomplished with the minimum of fuss and in record time.

And yet, on the face of it, one would have thought that it was a much more difficult thing to achieve than getting an owl to spread its wings. Simply, I wanted to get some shots of an egg-eating snake robbing a nest. Egg-eating snakes measure about two feet in length and are very slender. Coloured a pinkish-brown, mottled with darker markings, they have strange, protuberant eyes of a pale silvery colour with fine vertical pupils like a cat's. The curious thing about them is that about three inches from the throat (internally, of course) the vertebrae protrude, hanging down like stalactites. The reptile engulfs an egg, whole, and this passes down its body until it lies directly under these vertebrae. Then the snake contracts its muscles and the spikes penetrate the egg and break it; the yolk and white are absorbed and the broken shell, now a flattened pellet, is regurgitated. The whole process is quite extraordinary and had never, as far as I knew, been recorded on film.

We had, at that time, six egg-eating snakes, all of which were, to my delight, identical in size and colouration. The local children did a brisk trade in bringing us weaver birds' eggs to feed this troupe of reptiles with, for they seemed capable of eating any number we cared to put in their cage. In fact, the mere introduction of an egg into the cage changed them from a somnolent pile of snakes to a writhing bundle, each endeavouring to get at the egg first. But, although they behaved so beautifully in the cage, after my experiences with Woody and the water chevrotain I was inclined to be a bit pessimistic. However, I created a suitable set (a flowering bush, in the branches of which was placed a small nest) and collected a dozen small blue eggs as props. Then the snakes were kept without their normal quota of eggs for three days, to make sure they all had good appetites. This, incidentally, did not hurt them at all, for all snakes can endure considerable fasts which, in the case of some of the bigger constrictors, run into months or years. However, when my stars had got what I hoped was a good edge to their appetites, we started work.

The snakes' cage was carried out to the film set, five lovely blue eggs were placed in the nest, and then one of the reptiles was placed gently in the branches of the bush, just above the nest. I started the camera and waited.

The snake lay flaccidly across the branches, seeming a little dazed by the sunlight after the cool dimness of its box. Then its tongue started to flick in and out, and it turned its head from side to side in an interested manner. With smooth fluidity it started to trickle through the branches towards the nest. Slowly it drew closer, and when it reached the rim of the nest it peered over the edge and down at the eggs with its fierce silvery eyes. Its tongue flicked in and out as if it were smelling the eggs, and the snake nosed them gently like a dog with a pile of biscuits. Then it pulled itself a little farther into the nest, turned its head sideways, opened its mouth wide, and started to engulf one of the eggs. Now all snakes have jaws so constructed that they can dislocate the hinge; this enables them to swallow a prey that, at first sight, looks too big to pass through their mouths. The egg-eater neatly dislocated his jaws, and the skin of his throat stretched until each scale stood out individually and you could see the blue of the egg shining through the fine, taut skin as the egg was forced slowly down his throat. When the egg was about an inch down his body he paused for a moment's meditation and then swung himself out of the nest and into the branches. Here, as he made his way along, he rubbed the great swelling in his body against the branches so that the egg was forced farther and farther down.

Elated with this success, we returned the snake to his box so that he could digest his meal in comfort, and I shifted the camera's position and put on my big lenses for close-up work. We put another egg into the nest to replace the one taken, and then got out another egg-eater. This was the beauty of having all the snakes of the same size and colouration: as the first snake would now not look at another egg until he had digested the first, he could not be used in the close-up shots. But the new one was identical to look at and hungry as a hunter, and so without any trouble whatsoever I got all the close-up shots I needed as he glided rapidly down to the nest and took an egg. I did the whole thing all over again with two other snakes, and on the finished film these four separate sequences were intercut and no one, seeing the finished product, could tell that they were seeing four different snakes.

All the Bafutians, including the Fon, were fascinated by our filming activities, for not long before they had seen their very first cinema. A mobile motion-picture van had come out to Bafut and shown them a colour film of the Coronation, and they had been terribly thrilled with it. In fact it was still a subject of grave discussion when we were there, nearly a year and a half later. Thinking that the Fon and his council would be interested to learn more about filming, I invited them to come across one morning and attend a filming session and they accepted with delight.

"What are you going to film?" asked Jacquie.

"Well, it doesn't really matter, so long as it's innocuous," I said.

"Why innocuous?" asked Sophie.

"I don't want to take any risks. If I got the Fon bitten by something I would hardly be persona grata, would I?"

"Good God, no, that would never do," said Bob. "What sort of thing did you have in mind?"

"Well, I want to get some shots of those pouched rats, so we might as well use them. They can't hurt a fly."

So the following morning we got everything ready. The film set, representing a bit of forest floor, had been constructed on a Dexion stand, one of our specially made nylon tarpaulins had been erected nearby, under which the Fon and his court could sit, and under it were placed a table of drinks and some chairs. Then we sent a message over to the Fon that we were ready for him.

We watched him and his council approaching across the great courtyard, and they were a wonderful spectacle. First came the Fon, in handsome blue and white robes, his favourite wife trotting along beside him, shading him from the sun with an enormous orange and red umbrella. Behind him walked the council members in their fluttering robes of green, red, orange, scarlet, white, and yellow. Around this phalanx of colour some forty-odd of the Fon's children skipped and scuttled about like little black beetles round a huge, multicoloured caterpillar. Slowly the procession made its way round the rest house and arrived at our improvised film studio.

"Morning, my friend," called the Fon, grinning. "We done come for see dis cinema."

"Welcome, my friend," I replied. "You like first we go have a drink together?"

"Wah! Yes, I like," said the Fon, lowering himself cautiously into one of our camp chairs.

I poured out the drinks, and as we sipped them I explained the mysteries of motion-picture photography to the Fon, showing him how the camera worked and what the film itself looked like, and explaining how each little picture was of a separate movement.

"Dis filum you take, when we go see um?" asked the Fon, when he had mastered the basic principles of photography.

"I have to take um for my country before it get ready," I said sadly, "so I no fit show you until I go come back for the Cameroons."

"Ah, good," said the Fon, "so when you go come back for dis ma country we go have happy time and you go show me dis your filum."

We had another drink to celebrate the thought of my impending return to Bafut.

By this time everything was ready to show the Fon how one went about making a sequence. Sophie, as continuity girl, wearing trousers, shirt, sun glasses, and an outsize straw hat, was perched precariously on a small camp stool, her pad and pencil at the ready to make notes of each shot I took. Near her Jacquie, a battery of still cameras slung round her, was crouched by the side of the recording machine. Near the set Bob stood in the role of

dramatic coach, armed with a twig and the box in which
our stars were squeaking vociferously. I set up the camera,
took up my position behind it, and gave the signal for
action. The Fon and councillors watched silent and absorbed as Bob gently tipped the two pouched rats out on to
the set, and then guided them into the right positions with
his twig. I started the camera, and at the sound of its high-
pitched humming a chorus of appreciative "Ahs" ran
through the audience behind me. It was just at that moment that a small boy carrying a calabash wandered into
the compound and, oblivious of the crowd, walked up to
Bob and held up his offering. I was fully absorbed in
peering through the view-finder of the camera, and so I
paid little attention to the ensuing conversation that Bob
had with the child.

"Na whatee dis?" asked Bob, taking the calabash, which
had its neck plugged with green leaves.

"Beef," said the child succinctly.

Instead of inquiring more closely into the nature of the
beef, Bob pulled out the plug of leaves blocking the neck
of the calabash. The result surprised not only him but
everyone else as well. Six feet of agile and extremely angry
green mamba shot out of the interior of the calabash like
a Jack-in-the-box and fell to the ground.

"Mind your feet!" Bob shouted warningly.

I removed my eye from the view-finder of the camera
to be treated to the somewhat disturbing sight of the green
mamba sliding determinedly through the legs of the tri-
pod towards me. I leaped upwards and backwards with an
airy grace that only a prima ballerina treading heavily
on a tack could have emulated. Immediately pandemonium broke loose. The snake slid past me and made for
Sophie at considerable speed. Sophie took one look and
decided that discretion was the better part of valour. She

seized her pencil, pad, and, for some obscure reason, her camp stool as well, and ran like a hare towards the massed ranks of the councillors. Unfortunately this was the way the snake wanted to go as well, so he followed hotly on her trail. The councillors took one look at Sophie, apparently leading the snake into their midst, and they did not hesitate for a moment. As one man, they turned and fled. Only the Fon remained, rooted to his chair, so wedged behind the table of drinks that he could not move.

"Get a stick," I yelled to Bob and ran after the snake. I knew, of course, that the snake would not deliberately attack someone. It was merely trying to put the greatest possible distance between itself and us. But when you have fifty panic-stricken Africans, all barefoot, running madly in all directions, accompanied by a frightened and deadly snake, an accident is possible. The scene now, according to Jacquie, was fantastic. The council members were running across the compound, pursued by Sophie, who was pursued by the snake, who was pursued by me, who, in turn, was being pursued by Bob with a stick. The mamba had, to my relief, bypassed the Fon. The wave of battle having passed him by, the Fon sat there and did nothing more constructive than help himself to a quick drink.

At last Bob and I managed to corner the mamba against the rest-house steps. We held it down with a stick, picked it up, and popped it into one of our capacious snake bags. I returned to the Fon and found the council members drifting back from various points of the compass to join their monarch. Now, in any other part of the world if you had put to flight a cluster of dignitaries by introducing a snake into their midst, you would have had to put up with endless recriminations, sulks, wounded dignity, and other exhausting displays of human nature. But not so with the Africans. The Fon sat in his chair, beaming. The council-

lors chattered and laughed as they approached, clicking
their fingers at the danger that was past, making fun of
each other at having run so fast, and generally thoroughly
enjoying the humorous side of the situation.

"You done hold um, my friend?" asked the Fon, generously pouring me out a large dollop of my own whisky.

"Yes," I said, taking the drink gratefully, "we done hold
um."

The Fon leaned across and grinned at me mischievously.
"You see how all dis ma people run?" he asked.

"Yes, they run time no dere," I agreed.

"They de fear," explained the Fon.

"Yes. Na bad snake dat."

"Na true, na true," agreed the Fon, "all dis small small
man de fear dis snake too much."

"Yes."

"I never fear dis snake," said the Fon, "all dis ma people
dey de run. Dey de fear too much. But I never run."

"No, my friend, na true. You never run."

"I no de fear dis snake," said the Fon, in case I had
missed the point.

"Na true. But dis snake 'e de fear you."

" 'E de fear me?" asked the Fon, puzzled.

"Yes, dis snake no fit bite you. Na bad snake, but he no
fit kill Fon of Bafut."

The Fon laughed uproariously at this piece of blatant
flattery, and then, remembering the way his councillors
had fled, he laughed again, and the councillors joined
him. At length, still reeling with merriment at the incident, they left us, and we could hear their chatter and hi-
larious laughter long after they had disappeared. This is
the only occasion when I have know a green mamba to
pull off a diplomatic *coup d'état*.

MAIL BY HAND

My good friend,
  Good morning to you all. I received your note, but sorry my sickness is still going on as it was yesterday.
  I was sorry for I failed coming to drink with you, due to the sickness.
  I was grateful for the bottle of whisky and the medicine which you sent to me. I used the medicine yesterday evening and today morning, but no improvement yet. The thing which is giving me trouble is cough, if you can get some medicine for it, kindly send it to me through bearer.
  I think that whisky will also help, but I do not know yet. Please send me some gin if any.
  I am lying on bed.

Yours good friend,
Fon of Bafut

6: BEEF WITH HAND LIKE MAN

Of all the animals that one gets on a collecting trip the
ones that fascinate me most are, I think, members of the
monkey tribe. They are delightfully childlike, with
their quick intelligence, their uninhibited habits, their
rowdy, eager live-for-the-moment attitude towards life,
and their rather pathetic faith in you when they have accepted you as a foster parent.

In the Cameroons, monkeys are one of the staple items
of diet, and, as there are no enforced laws covering the
number that are shot or the season at which they are shot,
it is natural that a vast quantity of females carrying young
are slaughtered. The mother falls from the trees with the
baby still clinging tightly to her body, and in most cases
the infant is unhurt. Generally the baby is then killed and
eaten with the mother; occasionally the hunter will take
it back to his village, keep it until it is adult, and then eat

it. But when there is an animal collector in the vicinity, of course, all these orphans end up with him, for he is generally willing to pay much more than the market price for the living animal. So at the end of two or three months in a place like the Cameroons you generally find that you are playing foster parent to a host of monkeys of all shapes and ages.

In Bafut, towards the end of the trip, we had seventeen monkeys (not counting apes and the more primitive members of the tribe, such as pottos and bushbabies) and they caused us endless amusement. Probably the most colourful were the patas monkeys, slender creatures about the size of a terrier, with bright gingery red fur, soot-black faces, and white shirt fronts. In the wild state these monkeys live in the grasslands, rather than the forest, walking about like dogs in large family groups, assiduously searching the grass roots and rotten logs for insects or birds' nests, turning over stones to get worms, scorpions, spiders, and other delicacies. Periodically they will stand up on their hind legs to peer over the grass or, if the grass is too tall, they will leap straight up in the air as though on springs. Then if they see anything that smacks of danger they utter loud cries of "Proup . . . proup . . . proup," and canter away through the grass, with a swinging stride like little red race horses.

The four patases we had lived in a large cage together, and when they were not carefully grooming each other's fur with expressions of intense concentration on their sad, black faces, they were indulging in weird sorts of Oriental dances. Patases are the only monkeys I know of that really do dance. Most monkeys will, during an exuberant game, twirl round or jump up and down, but patases have worked out special dance sequences for themselves and, moreover, have quite an extensive repertoire. They would start by bouncing up and down on all fours like a rubber ball, all four feet leaving the ground simultaneously, going faster and faster and higher and higher, until they were leaping almost two feet into the air. Then they would stop and start a new series of "steps." Keeping their back legs and hind quarters quite still, they would start to swing the fronts of their bodies from side to side like a pendulum, twisting their heads from left to right as they did so. When they had done this twenty or thirty times they would start a new variation. This consisted of standing up stiff and straight on their hind legs, arms stretched above their heads, faces peering up at the roof of their cage, and then staggering round and round in circles until they were so dizzy that they fell backwards. This whole dance would be accompanied by a little song, the lyrics of which went like this: "Waaaaow . . . waaaaow . . . proup . . . proup . . . waaaaow . . . proup," which was considerably more attractive and comprehensible than the average popular song sung by the average popular crooner.

The patases, of course, adored live food of any description and felt that their day was incomplete if they did not have a handful of grasshoppers, or some birds' eggs, or a brace of juicy, hairy spiders apiece. But the thing that was to them the caviar of life were the larvae of the palm beetles. Palm beetles are oval-shaped insects about two inches in length, which are very common in the Cameroons. They lay their eggs in rotting tree trunks, but show a marked preference for the soft, fibrous interior of the palm trees. Here, in a moist, soft bed of food, the egg hatches out and the grub soon grows into a livid white, maggot-like creature about three inches long and as thick as your thumb. These fat, twitching grubs were considered by the patases to be the food of the gods, and the shrieks of delight that would greet my appearance with a tinful of them would be almost deafening. The curious thing was that, although they adored eating the larvae, they were really scared of them. I would empty the grubs on the floor of the cage and the patases would squat round the pile, still screaming with pleasure, and keep touching the delicacies with trembling, tentative fingers; if the grubs moved, they would hastily withdraw their hands and wipe them hurriedly on their fur. At last one of them would grab a fat larva and, screwing up his face and closing his eyes tightly, he would stuff the end into his mouth and bite hard. The larva, of course, would respond to this unkind decapitation with a frantic dying wriggle and the patases would drop it hurriedly on the floor, wipe his hands and, still sitting with tightly closed eyes and screwed-up face, would munch on the morsel he had bitten off. They reminded me of debutantes being introduced to their first fresh oysters.

Unwittingly one day—under the impression that I was doing them a kindness—I caused pandemonium in the patases' cage. An army of local children kept us supplied with live food for the animals, and they would arrive just after dawn with calabashes full of snails, birds' eggs, beetle larvae, grasshoppers, spiders, tiny hairless rats, and other strange food that our animals enjoyed. On this particular morning one lad had brought in, as well as his normal offering of snails and palm-beetle larvae, the larvae of two Goliath beetles. Now Goliath beetles are the biggest beetles in the world, an adult measuring some six inches in length and four inches across the back, so it goes without saying that the larvae were monsters. They were about six inches long, and as thick as my wrist. They were the same horrid unhealthy white as the palm-beetle larvae, but they were much fatter, their skin being wrinkled and folded and tucked like an eiderdown. They had flat, nut-brown heads the size of a shilling with great curved jaws that could give you quite a pinch if you handled them incautiously. I was very pleased with these monstrous, bloated, maggot-like creatures, for I felt that, since the patases liked palm-beetle larvae so much, their delight would know no bounds when they set eyes on these gigantic titbits. So I put the Goliath larvae in the usual tin with the other grubs and went to give them to the patases as a light snack before they had their breakfast.

As soon as they saw the familiar tin on the horizon the patases started to dance up and down excitedly, crying, "Proup . . . proup," and then as I was opening the door they sat down in a circle, their little black faces wearing a worried expression, their hands held out beseechingly. I pushed the tin through the door and tipped it up so that the two Goliath larvae fell onto the floor of the cage with a soggy thud, where they lay unmoving. To say that the patases were surprised is an understatement; they uttered faint squeaks of astonishment and shuffled backwards on their bottoms, surveying the barrage-balloon-like larvae with a certain horrified mistrust. They watched them narrowly for a minute or so, but as there was no sign of movement from the larvae, they gradually became braver, and shuffled closer to examine this curious phenomena more minutely. Then, having studied the grubs from every possible angle, one of the monkeys, greatly daring, put out a hand and prodded a grub with a tentative forefinger. The grub, who up till then had been lying on his back in a sort of trance, woke up, gave a convulsive wriggle, and rolled over majestically onto his tummy. The effect of this movement on the patases was tremendous. Uttering wild screams of fear they fled in a body to the farthest corner of the cage, where they indulged in a disgracefully cowardly

scrimmage, vaguely reminiscent of the Eton wall game, each one doing his best to get into the extreme corner of the cage, behind all his companions. Then the grub, having pondered for a few seconds, started laboriously to drag his bloated body across the floor towards them, the patases showed such symptoms of collective hysteria that I was forced to interfere and remove the grubs. I put them in Ticky the black-footed mongoose's cage and she, who was not afraid of anything, disposed of them in four snaps and two gulps. But the poor patases were in a twittering state of nerves for the rest of the day; ever after that, when they saw me coming with the beetle-larvae tin, they would retreat hurriedly to the back of the cage until they were sure that the tin contained nothing more harmful or horrifying than palm-beetle larvae.

One of our favourite characters in the monkey collection was a half-grown female baboon called Georgina. She was a creature of tremendous personality and a wicked sense of humour. She had been hand-reared by an African who had kept her as a sort of pet-cum-watchdog, and we had purchased her for the magnificent sum of ten shillings. Georgina was, of course, perfectly tame, so she wore a belt round her waist, to which was attached a long rope, and every day she was taken out and tied to one of the trees in the compound below the rest house. For the first couple of days we tied her up fairly near the gate leading into the compound, through which came a steady stream of hunters, old ladies selling eggs, and hordes of children with insects and snails for sale. We thought that this constant procession of humanity would keep Georgina occupied and amused. It certainly did, but not in the way we intended. She very soon discovered that she could go to the end of her rope and crouch out of sight behind the hibiscus hedge, very near the gate. Then, when some unsuspecting

African came into the compound she would leap out of her ambush and embrace him round the legs, at the same time uttering such a blood-curdling scream as to make even the staunchest nerves falter and break.

Her first successful ambuscade was perpetrated on an old hunter who, clad in his best robe, was bringing a calabash full of rats to us. He had approached the rest house slowly and with great dignity, as befitted one who was bringing such rare creatures for sale, but his aristocratic pose was rudely shattered as he came through the gate. Feeling his legs seized in Georgina's iron grasp, and hearing her terrifying scream, he dropped his calabash of rats, which promptly broke so that they all escaped; he leaped straight up in the air with a roar of fright and then fled down the road in a very undignified manner and at a speed quite remarkable for one of his age. It cost me three packets of cigarettes and considerable tact to soothe his ruffled feelings. Georgina, meanwhile, sat there looking as if butter would not melt in her mouth and, as I scolded her, merely raised her eyebrows, displaying her pale pinkish eyelids in an expression of innocent astonishment.

Her next victim was a handsome sixteen-year-old girl who had brought a calabash full of snails. The girl, however, was almost as quick in her reactions as Georgina. She saw the baboon out of the corner of her eye, just as Georgina made her leap. The girl sprang away with a squeak of fear and Georgina, instead of getting a grip on her legs, merely managed to fasten onto the trailing corner of her sarong. The baboon gave a sharp tug and the sarong came away in her hairy paw, leaving the unfortunate damsel quite naked. Georgina, screaming with excitement, immediately put the sarong over her head like a shawl and sat chattering happily to herself, while the poor girl, overcome with embarrassment, backed into the hibiscus hedge, endeavouring to cover all the vital portions of her anatomy with her hands. Bob, who happened to witness this incident with me, needed no encouragement to volunteer to go down into the compound, retrieve the sarong, and return it to the damsel.

So far Georgina had had the best of these skirmishes, but the next morning she overplayed her hand. A dear old lady, weighing some two hundred pounds, came waddling and wheezing up to the rest-house gate, balancing carefully on her head a kerosene tin full of groundnut oil, which she was hoping to sell to Phillip the cook. Phillip, having spotted the old lady, rushed out of the kitchen to warn her about Georgina, but he arrived on the scene too late. Georgina leaped from behind the hedge with the stealth of a leopard and threw her arms round the old lady's fat legs, uttering her frightening war cry as she did so. The poor old lady was far too fat to jump and run as the other victims had done, so she remained rooted to the spot, uttering screams that for quality and quantity closely rivalled the sounds Georgina was producing. While they indulged in this cacophonous duet, the kerosene tin wobbled precariously on the old lady's head. Phillip came clumping across the compound on his enormous feet, roaring hoarse instructions to the old lady, none of which she appeared to hear or obey. When he reached the scene of the battle, Phillip in his excitement did a very silly thing. Instead of confining his attention to the tin on the old lady's head, he concentrated on her other end, and seizing Georgina attempted to pull her away. Georgina, however, was not going to be deprived of such a plump and prosperous victim so easily and, screaming indignantly, she clung on like a limpet. Phillip, holding the baboon round the waist, tugged with all his might. The old lady's vast bulk quivered like a mighty tree on the point of falling and the kerosene tin on her head gave up the unequal struggle with the laws of gravity and fell to the ground with a crash. A wave of oil leaped into the air as the tin struck the ground and covered the three protagonists in a golden, glutinous waterfall. Georgina, startled by this new, cowardly, and possibly dangerous form of warfare, gave a grunt of fright, let go the old lady's legs, and retreated to the full stretch of her rope, where she sat down and endeavoured to rid her fur of the sticky oil. Phillip stood looking as though he were slowly melting from the waist down, and the front of the old lady's sarong was equally sodden.

"Wah!" roared Phillip, ferociously. "You *stupid* woman, why you throw dis oil for ground?"

"Foolish man," screamed the old lady, equally indignant, "dis beef come for bite me, how I go do?"

"Dis monkey no go bite you, blurry fool, na tame one," roared Phillip, "and now look dis my clothes done spoil. Na your fault dis."

"No be my fault, no be my fault," screeched the old lady, her impressive bulk quivering like a dusky volcano. "Na your own fault, bushman, an' all my dress done spoil, all dis my oil done throw for ground."

"Blurry foolish woman," blared Phillip, "you be bushwoman, you done throw dis oil for ground for no cause. All dis my clothes done ruin."

He stamped his large foot in irritation, with the unfortunate result that it landed in the pool of oil and splashed over the front of the old lady's already dripping sarong. Giving a scream like a descending bomb, the old lady stood there quivering as if she would burst. Then she found her voice. She uttered only one word, but I knew then was the time to interfere. "Ibo!" she said malevolently.

Phillip reeled before this insult. The Ibos are a Nigerian

tribe whom the Cameroonians regard with horror and loathing, and to call someone in the Cameroons an Ibo is the deadliest insult you can offer. Before Phillip could collect his wits and do something violent to the old lady, I intervened, trying to control my laughter. I soothed the good lady, gave her compensation for her sarong and lost oil, and then somewhat mollified the still simmering Phillip by promising to give him a new pair of shorts, socks, and a shirt out of my own wardrobe. Then I untied the glutinous Georgina, and removed her to a place where she could not perform any more expensive attacks on the local population.

But Georgina had not finished yet. Unfortunately I tied her up under the lower veranda, close to a room which we used as a bathroom. In it was a large, circular, red plastic bowl which was prepared each evening so that we could wash the sweat and dirt of the day's work from our bodies. The difficulty of bathing in this plastic bowl was that it was a shade too small. In order to recline in the warm water and enjoy it you were forced to leave your feet and legs outside, as it were, resting on a wooden box. The bowl being slippery, a considerable effort was generally required to get up from this reclining position in order to get the soap or the towel or some other necessity. This bath was not, in my opinion, the most comfortable in the world, but it was the best we could do in the circumstances.

Sophie adored her bath, and would spend far longer than anyone else over it, lying back luxuriously in the warm water, smoking a cigarette, and reading a book by the light of a tiny hurricane lantern. On this particular night her ablutions were not so prolonged. The battle of the bathroom commenced when one of the staff came to her and said, in the conspiratorial manner they always seemed to adopt, "Barf ready, madam." Sophie got her

book and her tin of cigarettes and wandered down to the
bathroom. She found it already occupied by Georgina,
who had discovered that the length of her rope and the
position in which I had tied her allowed her access to this
interesting room. She was sitting by the bath, dipping the
towel in the water and uttering little throaty cries of satisfaction. Sophie shooed her out, called for a new towel, and
then, after closing the door, she undressed and lowered
herself into the warm water.

Unfortunately, as she soon discovered, Sophie had not
shut the door properly. Georgina had never seen anyone
bathe before and she was not going to let such a unique
opportunity pass without taking full advantage of it. She
hurled herself against the door and threw it wide open.
Sophie now found herself in a predicament; she was so
tightly wedged in the bath that she could not get out and
shut the door without considerable difficulty, and yet to
lie there with the door open was not possible. With a
great effort she leaned out of the bath and reached for her
clothes, which she had fortunately placed nearby. Georgina, seeing this, decided that it was the beginning of a
promising game; she jumped forward, clasped Sophie's
raiment to her hairy bosom, and ran outside with it. This
left only the towel. Sophie struggled out of the bath,
draped herself in this inadequate covering, and, having
made sure that no one was around, went outside to try to
retrieve her garments.

Georgina, seeing Sophie was entering into the spirit of
the game, gave a chattering cry of delight and as Sophie
made a dart at her she ran back into the bathroom and hurriedly put Sophie's clothes in the bath. Taking Sophie's
cry of horror to be encouragement, she then seized the tin
of cigarettes and put that in the bath too, presumably to
see if it would float. It sank, and forty-odd cigarettes

floated dismally to the surface. Then, in order to leave no stone unturned in her efforts to give Sophie pleasure, Georgina tipped all the water out of the bath. Attracted by the uproar, I appeared on the scene just in time to see Georgina leap nimbly into the bath and start to jump up and down on the mass of sodden cigarettes and clothes, rather after the manner of a wine-treader. It took considerable time to remove the excited baboon and get Sophie fresh bath water, cigarettes, and clothing, by which time the dinner was cold. So Georgina was responsible for a really exhilarating evening.

But of all our monkey family it was the apes, I think, that gave us the most pleasure and amusement. The first one we got was a baby male chimpanzee who arrived one morning, reclining in the arms of a hunter, with such an expression of sneering aristocracy on his small, wrinkled face that one got the impression he was employing the hunter to carry him about, in the manner of an Eastern potentate. He sat quietly on the rest-house steps, watching us with intelligent, scornful brown eyes, while the hunter and I bargained over him, rather as though this sordid wrangling over money were acutely distasteful to a chimpanzee of his upbringing and background. When the bargain had been struck and the filthy lucre had changed hands, this simian aristocrat took my hand condescendingly and walked into our living room, peering about him with an air of ill-concealed disgust, like a duke visiting the kitchen of a sick retainer, determined to be democratic however unsavoury the task. He sat on the table and accepted our humble offering of a banana with the air of one who is weary of the honours that have been bestowed upon him throughout life. Then and there we decided that he must have a name befitting such a blue-blooded primate, so we christened him Cholmondely St. John, pronounced, of course, Chumley Sinjun. Later, of course, when we got to know him better, he allowed us to become quite familiar with him and call him Chum; or sometimes, in moments of stress, "you bloody ape," but this latter term always made us feel as though we were committing *lèse-majesté*.

We built Cholmondely a cage (to which he took grave exception) and allowed him out only at set times during the day, when we could keep an eye on him. First thing in the morning, for example, he was let out of his cage, and accompanied a member of the staff into our bedroom with the morning tea. He would gallop across the floor and leap into bed with me, give me a wet and hurried kiss as greeting, and then, with grunts and staccato cries of "Ah! Ah!" would watch the tea tray put in position and examine it carefully to make sure that his cup (a large tin one for durability) was there. He would sit back and watch me carefully while I put milk, tea, and sugar (five spoons) into his mug, and then take it from me with twitching, excited hands, bury his face in it, and drink with a noise like a very large bath running out. He would not even pause for breath, but would lift the mug higher and higher, until it was upside down over his face; then there was a long pause as he waited for the delicious, semi-melted sugar to slide down into his open mouth. Having made quite sure that there was no sugar left at the bottom, he would sigh deeply, belch in a reflective manner, and hand the mug back to me in the vague hope that I would refill it. Seeing that this wish was not going to be fulfilled, he would watch me drink my tea, and then set about the task of entertaining me.

There were several games he had invented for my benefit, and all were exhausting to take part in at that hour of the morning. To begin with, he would prowl down to the end of the bed and squat there, giving me surreptitious glances to make sure I was watching. Then he would insert a cold hand under the bedclothes and grab my toes. I was supposed to lean forward with a roar of pretended rage, and he would leap off the bed and run to the other side of the room, watching me over his shoulder with a wicked expression of delight in his brown eyes. When I tired of this game I would pretend to be asleep, and he would walk slowly and cautiously down the bed and peer into my face for a few seconds. Then he would shoot out a long arm, pull a handful of my hair, and rush down to the bottom of the bed before I could catch him. If I did succeed in grabbing him I would put my hands round his neck and tickle his collar bones, while he wriggled and squirmed, opening his mouth wide and drawing back his lips to display a vast acreage of pink gums and white teeth, giggling hysterically like a child.

Our second acquisition was a large five-year-old chimp called Minnie. A Dutch farmer turned up one day and said that he was willing to sell us Minnie, as he was soon due to go on leave and did not want to leave the animal to the mercies of his staff. We could buy Minnie if we went and fetched her. The Dutchman's farm was some fifty miles away at a place called Santa, so we arranged to go there in the Fon's Land-Rover, see the chimp, and, if she proved healthy, buy her, and bring her back to Bafut. So, taking a large crate with us, we set off very early one morning, thinking we would be back with the chimp in time for a late lunch.

To get to Santa we had to drive out of the valley in which Bafut lay, climb the great Bemenda escarpment (an almost sheer three-hundred-foot cliff), and drive on into the range of mountains that lay beyond it. The landscape was white with heavy morning mist that, waiting for the

sun to rise and drag it into the sky in great toppling columns, lay placidly in the valleys like pools of milk, out of
which appeared the peaks of hills and escarpments like
strange islands in a pallid sea. As we drove higher into
the mountains we went more slowly, for here the slight
dawn wind, in frail spasmodic gusts, rolled and pushed
these great banks of mist so that they swirled and poured
themselves across the road like enormous pale amoebas,
and you would suddenly round a corner and find yourself
deep in the belly of a mist bank, visibility cut down to a
few yards. At one point, as we edged our way through a
bank of mist, there appeared in front of us what seemed,
at first sight, to be a pair of elephant tusks. We shuddered
to a halt, and out of the mist loomed a herd of long-horned
Fulani cattle, which surrounded us in a tight wedge, peering interestedly through the Land-Rover windows. They
were huge, beautiful beasts of a dark chocolate brown,
with enormous melting eyes and a massive spread of white
horns, sometimes as much as five feet from tip to tip. They
pressed closely around us, their warm breath pouring from
their nostrils in white clouds, the sweet cattle smell of their
bodies heavy in the cold air, while the guide cow's bell
tinkled pleasantly with each movement of her head. We
sat and surveyed each other for a few minutes and then
there was a sharp whistle and a harsh cry as the herdsman
appeared out of the mist, a typical Fulani, tall and slender, with fine-boned features and a straight nose, looking
like an ancient Egyptian mural.

"*Iseeya,* my friend," I called.

"Morning, Masa," he answered, grinning and slapping
the dewy flank of an enormous cow.

"Na your cow dis?"

"Yes, sah, na ma own."

"Which side you take um?"

"For Bemenda, sah, for market."

"You fit move um so we go pass?"

"Yes, sah, yes sah, I go move um." He grinned and with loud shouts he urged the cows onwards into the mist, dancing from one to the other and beating a light tattoo on their flanks with his bamboo walking stick. The great beasts moved off into the mist, giving deep contented bellows, the guide cow's bell tinkling pleasantly.

"Thank you, my friend, walker good," I called after the tall herdsman.

"Tank you Masa, tank you," came his voice out of the mist, against a background of deep, bassoon-like cow calls.

By the time we reached Santa the sun was up and the mountains had changed to golden green, their flanks still striped here and there with tenacious streaks of mist. We reached the Dutchman's house to find that he had been unexpectedly called away. However, Minnie was there and it was she that we had come for. She lived, we discovered, in a large circular enclosure that the Dutchman had built for her, surrounded by a tallish wall and furnished simply but effectively with four dead trees, planted upright in cement, and a small wooden house with a swing door. You gained access to this enclosure by lowering a form of drawbridge in the wall which allowed you to cross the dry moat that surrounded Minnie's abode.

Minnie was a large, well-built chimp about three feet six in height, and she sat in the branches of one of her trees and surveyed us with an amiable if slightly vacuous expression. We surveyed each other silently for some ten minutes, while I endeavoured to assess her personality. Although the Dutchman had assured me that she was perfectly tame, I had had enough experience to know that even the tamest chimp, if it takes a dislike to you, can be a

nasty creature to have a rough-and-tumble with, and Minnie, though not very tall, had an impressive bulk.

Presently I lowered the drawbridge and went into the
enclosure, armed with a large bunch of bananas with
which I hoped I could purchase my escape should my esti-
mation of her character be faulty. I sat on the ground, the
bananas on my lap, and waited for Minnie to make the
first overtures. She sat in the tree watching me, thoughtfully slapping her rotund tummy with her large hands.
Then, having decided that I was harmless, she climbed
down from the tree and loped over to where I sat. She
squatted down about a yard away and then held out a
hand to me. Solemnly I shook it. Then I, in turn, held out
a banana, which she accepted and ate with small grunts of
satisfaction.

Within half an hour she had eaten all the bananas and
we had established some sort of friendship; that is to say,
we played pat-a-cake, we chased each other round her compound and in and out of her hut, and we climbed one of
the trees together. At this point I thought it was a suitable
moment to introduce the crate into the compound. We
carried it in, placed it on the grass with its lid nearby, and
allowed Minnie plenty of time to examine it and decide it
was harmless. The problem was to get Minnie into the
crate, first, without frightening her too much and, second,
without getting bitten. As she had never in her life been
confined in a box or small cage, I could see that the whole
thing presented difficulties, especially as her owner was
not there to lend his authority to the manœuvre.

So, for three and a half hours I endeavoured, by example, to show Minnie that the crate was harmless. I sat in
it, lay in it, jumped about on top of it, even crawled round
with it on my back like a curiously shaped tortoise. Minnie

enjoyed my efforts to amuse her immensely, but she still treated the crate with a certain reserve. The trouble was that I knew I should have only one opportunity to trap her, for if I messed it up the first time and she realized what I was trying to do, no amount of coaxing or cajoling would induce her to come anywhere near the crate. Slowly she had to be lured to the crate so that I could tip it over on top of her. So, after another three quarters of an hour of concentrated and exhausting effort, I got her to sit in front of the upturned crate and take bananas from inside it. Then came the great moment.

I baited the box with a particularly succulent bunch of bananas and then sat myself behind it, eating a banana myself and looking around the landscape nonchalantly, as though nothing could be farther from my mind than the thought of trapping chimpanzees. Minnie edged forward, darting surreptitious glances at me. Presently she was squatting close by the box, examining the bananas with greedy eyes. She gave me a quick glance and then, as I seemed preoccupied with my fruit, she leaned forward and her head and shoulders disappeared inside the crate. I hurled my weight against the back of the box, so that it toppled over her, and then jumped up and sat heavily on top so that she could not bounce it off. Bob rushed into the compound and added his weight and then, with infinite caution, we edged the lid underneath the crate, turned the whole thing over, and nailed the lid in place, while Minnie sat, surveying me malevolently through a knot hole and crying, "Ooo . . . oooo . . . oooo," plaintively, as if shocked to the core at my perfidy. After wiping the sweat from my face and lighting a much-needed cigarette, I glanced at my watch. It had taken four and a quarter hours to catch Minnie; I reflected that it could not have

taken much longer if she had been a wild chimpanzee leaping about in the forest. Tiredly we loaded her onto the
Land-Rover and set out for Bafut again.

At Bafut we had already constructed a large cage out of
Dexion for Minnie. It was not, of course, anywhere near as
big as what she was used to, but big enough so that she
would not feel too confined to begin with. Later she would
have to go into quite a small crate for the voyage home,
but after all the freedom she had been used to I wanted
to break her gradually to the idea of being closely confined. When we put her into her new cage she explored it
thoroughly with grunts of approval, banging the wire with
her hands and swinging on the perches to see how strong
they were. Then we gave her a big box of mixed fruit and
a large white plastic bowl full of milk, which she greeted
with hoots of delight.

The Fon had been very interested to hear that we were
getting Minnie, for he had never seen a large, live chimpanzee before. So that evening I sent him a note inviting
him to come over for a drink and to see the ape. He arrived just after dark, wearing a green and purple robe,
accompanied by six council members and his two favourite wives. After the greetings were over and we had ex-
changed small talk over the first drink of the evening, I
took the pressure lamp and led the Fon and his retinue
down the veranda to Minnie's cage, which, at first sight,
appeared to be empty. I lifted the lamp higher and we discovered that Minnie was in bed. She had made a nice pile
of dry banana leaves at one end of the cage and had settled
down in this, lying on her side, her cheek pillowed on one
hand, with an old sack we had given her carefully draped
over her body and tucked under her armpits.

"Wah!" said the Fon. " 'E sleep like man."

"Yes, yes," chorused the council members, " 'e sleep like man."

Minnie, disturbed by the lamplight and the voices, opened one eye to see what the disturbance was about. Seeing the Fon and his party, she decided that they might well repay closer investigation, so she threw back her sacking cover carefully and waddled over to the wire.

"Wah!" said the Fon, " 'e same same for man, dis beef."

Minnie looked the Fon up and down, and then decided that he might be inveigled into playing with her, so she beat a loud tattoo on the wire with her big hands.

The Fon and his party retreated hurriedly.

"No de fear," I said. "Na funning dis."

The Fon approached cautiously, an expression of astonished delight on his face. Cautiously he leaned forward and banged on the wire with the palm of his hand. Minnie, delighted, answered him with a fusilade of bangs that made him jump back and then crow with laughter.

"Look 'e hand, look 'e hand," he gasped. " 'E get hand like man."

"Yes, yes, 'e get hand same same for man," said the councillors.

The Fon leaned down and banged on the wire again, and Minnie once more responded.

"She play musica with you," I said.

"Yes, yes, na chimpanzee musica dis," said the Fon, and went off into peals of laughter. Greatly excited by her success, Minnie ran round the cage two or three times and did a couple of backward somersaults on her perches. Then she came and sat in the front of the cage, seized her plastic milk bowl, and placed it on her head, where it perched, looking incongruously like a steel helmet. The roar of laughter that this manœuvre provoked from the Fon and his councillors and wives caused half the village dogs to start barking.

" 'E get hat, 'e get hat," gasped the Fon, doubling up with mirth.

Feeling that it was going to be almost impossible to drag the Fon away from Minnie, I called for the table, chairs, and drinks to be brought out and placed on the veranda near the chimp's cage. So for half an hour the Fon sat there alternately sipping his drink and spluttering with laughter, while Minnie showed off like a veteran circus performer. Eventually, feeling somewhat tired by her performance, Minnie came and sat near the wire by the Fon, watching him as he drank with great interest, still wearing her plastic bowl helmet. The Fon beamed down at her. Then he leaned forward until his face was only some six inches away from Minnie's and lifted his glass.

"Shin-shin?" said the Fon.

To my complete astonishment Minnie responded by protruding her long, mobile lips and giving a prolonged raspberry of the juiciest variety.

The Fon laughed so loud and so long at this witticism that at last we were all in a state of hysterical mirth just watching him enjoy the jest. At length, taking a grip on himself, he wiped his eyes, leaned forward, and blew a raspberry at Minnie. But his was a feeble amateur effort compared to the one with which Minnie responded, which echoed up and down the veranda like a machine gun. So for the next five minutes—until the Fon had to give up because he was laughing so much, and out of breath—he and Minnie kept up a rapid crossfire of raspberries. Minnie was definitely the winner, judged by quality and quantity; also she had better breath control, so that her efforts were much more prolonged and sonorous than the Fon's.

At length the Fon left us, and we watched him walking

back across the great compound, occasionally blowing raspberries at his councillors, whereupon they would all double up with laughter. Minnie, with the air of a society
hostess after an exhausting dinner party, yawned loudly
and then went over and lay down on her banana-leaf bed,
covered herself carefully with the sack, put her cheek on
her hand, and went to sleep. Presently her snores rever-
berated along the veranda almost as loudly as her raspberries.

PART THREE:
COASTWARDS AND ZOOWARDS

MAIL BY HAND

Sir,

I have the honour most respectfully beg to submit this letter to you stating as follows:—

(1) I regret extremely at your leaving me, though not for bad but for good.

(2) At this juncture, I humbly and respectfully beg that you as my kind master should leave a good record of recommendation about me which will enable your successor to know all about me.

(3) Though I have worked with several Masters I have highly appreciated your ways then all.

Therefore should the Master leave some footprints behind on my behalf, I shall price that above all my dukedoms.

I have the honour to be, Sir,

Your obedient Servent,

Phillip Onaga (Cook)

7: A ZOO IN OUR LUGGAGE

It was time for us to start making preparations to leave Bafut and travel the three hundred-odd miles down to the coast. But there was a lot to be done before we could set out on the journey. In many ways this is the most harassing and dangerous part of a collecting trip. To begin with, to load your animals onto trucks and take them that distance, over roads that resemble a tank-training ground more than anything else, is in itself a major undertaking. But there are many other vital things to arrange as well. Your food supply for the voyage must be waiting for you at the port, and here again you cannot afford to make any mistakes, for you cannot take two hundred and fifty animals on board a ship for three weeks unless you have an adequate supply of food. All your cages have to be carefully inspected and any defects caused by six months' wear and tear have to be made good, because you cannot risk having an escape on board ship. So cages have to be rewired, new fastenings fixed on doors, new bottoms fitted onto cages that show signs of deterioration, and a hundred and one other minor jobs have to be done.

Taking all this into consideration, it is not surprising that you have to start making preparations for your departure sometimes a month before you actually leave your base camp for the coast. Everything, it seems, conspires against you. The local population, horrified at the imminent loss of such a wonderful source of revenue, redouble their hunting efforts so as to make the maximum profit before you leave, and this means that you are not only renovating old cages but constructing new ones as fast as you can to cope with this sudden influx of creatures. The local telegraph operator undergoes what appears to be a mental breakdown, so that the vital telegrams you send and receive are incomprehensible to both you and the recipient. When you are waiting anxiously for news of your food supplies for the voyage it is not soothing to the nerves to receive a telegram which states, in what is apparently Esperanto, "Message replied regret cannotob vary green balas well half pipe do?" which, after considerable trouble and expense, you get translated as, "Message received regret cannot obtain very green bananas will half ripe do?"

Needless to say, the animals soon become aware that something is in the wind and try to soothe your nerves in their own particular way: those that are sick get sicker, and look at you in such a frail and anæmic way that you are quite sure they will never survive the journey down to the coast; all the rarest and most irreplaceable specimens try to escape, and if successful hang around taunting you with their presence and making you waste valuable time in trying to catch them again; animals that had, up till then, refused to live unless supplied with special food, whether avocado pear or sweet potato, suddenly decide that they do not like this particular food any more, so frantic telegrams have to be sent cancelling the vast quantities of the delicacies you had just ordered for the voyage. Altogether this part of a collecting trip is very harassing.

The fact that we were worried and jumpy, of course, made all of us do silly things that only added to the confusion. The case of the clawed toads is an example of what I mean. Anyone might be pardoned for thinking that clawed toads were frogs at first glance. They are smallish creatures with blunt, froglike heads and smooth slippery skins which are most untoadlike. Also, they are almost completely aquatic, another untoadlike characteristic. To my mind they are rather dull creatures who spend ninety per cent of their time floating in the water in various abandoned attitudes, occasionally shooting to the surface to take a quick gulp of air. But, for some reason which I could never ascertain, Bob was inordinately proud of these wretched toads. We had two hundred and fifty of them and we kept them in a gigantic plastic bath on the veranda. Whenever Bob was missing, one was almost sure to find him crouched over this great cauldron of wriggling toads, an expression of pride on his face. Then came the day of the great tragedy.

The rainy season had just started and the brilliant sunshine of each day was being interrupted by heavy downpours of rain, which lasted only an hour or so, but during which the quantity of water that fell was quite prodigious. On this particular morning Bob had been crooning over his clawed toads, and when it started to rain he thought that they would be grateful if he put their bowl out in it. So he carefully carried the toads' bowl down the veranda and placed it on the top step, brilliantly positioned so that it received not only the rain itself but all the water that ran off the roof. Then he went away to do something

else and forgot all about it. The rain continued to rain as if determined to uphold the Cameroons' reputation for being one of the wettest places on earth, and gradually the bowl filled up. As the water level rose, the toads rose with it until they were peering over the plastic rim. Another ten minutes of rain and, whether they wanted to or not, they were swept out of the bowl by the overflow.

My attention was drawn to this instructive sight by Bob's moan of anguish when he discovered the catastrophe—a long-drawn howl of emotion that brought us all running from wherever we were. On the top step stood the plastic bowl, now completely devoid of toads. From it the water gushed down the steps, carrying with it Bob's precious amphibians. The steps were black with toads, slithering, hopping, and rolling over and over in the water. In amongst this Niagara of amphibians Bob, with a wild look in his eye, was leaping to and fro like an excited heron, picking up toads as fast as he could. Picking up a clawed toad is quite a feat. It's almost as difficult as trying to pick up a drop of quicksilver; apart from the fact that their bodies are incredibly slippery, the toads are very strong for their size and kick and wriggle with surprising energy. In addition, their hind legs are armed with small, sharp claws, and when they kick out with their muscular hind legs they are quite capable of inflicting a painful scratch. Bob, alternately moaning and cursing in anguish, was not in the calm, collected mood that is necessary for clawed-toad catching, so every time he had scooped up a handful of the creatures and was bounding up the steps to return them to their bath, they would squeeze from between his fingers and fall back onto the steps, to be immediately swept downwards again by the water. In the end it took five of us three quarters of an hour to collect all the toads and put them back in their bowl, and just as we had finished and were soaked to the skin it stopped raining.

"If you must release two hundred and fifty specimens you might at least choose a fine day and an animal that is reasonably easy to pick up," I said to Bob bitterly.

"I can't think what made me do such a silly thing," said Bob, peering dismally into the bowl, in which the toads, exhausted after their romp, hung suspended in the water, peering up at us in their normal popeyed, vacant way. "I do hope they're not damaged in any way."

"Oh, never mind about us. We can all get pneumonia galloping about in the rain, just as long as those repulsive little devils are all right. Would you like to take their temperatures?"

"You know," said Bob frowning, and ignoring my sarcasm, "I'm sure we've lost quite a lot. There doesn't seem to be anything like the number we had before."

"Well, I for one am not going to help you count them.

I've been scratched enough by clawed toads to last me a lifetime. Why don't you go and change and leave them alone? If you start counting them you'll only have the whole damn lot out again."

"Yes," said Bob, sighing, "I suppose you're right."

Half an hour later I let Cholmondely St. John, the chimp, out of his cage for his morning exercise, and stupidly took my eye off him for ten minutes. As soon as I heard Bob's yell, the cry of a mind driven past breaking point, I took a hasty look round and, not seeing Cholmondely St. John, I knew he was the cause of Bob's banshee wail. I hurried out to the veranda and found Bob wringing his hands in despair, while on the top step sat Cholmondely, looking so innocent that you could almost see his halo gleaming. Halfway down the steps, upside down, was the plastic bowl, and the steps below it and the compound beyond were freckled with hopping, hurrying toads.

We slithered and slipped in the red mud of the compound for an hour before the last toad was caught and put in the bowl. Then, breathing hard, Bob picked it up and in silence we made our way back to the veranda. As we reached the top step Bob's muddy shoes slipped under him and he fell, and the bowl rolled to the bottom, and for the third time the clawed toads set off joyfully into the wide world.

Cholmondely St. John was responsible for another escape, but this was less strenuous and more interesting than the clawed-toad incident. In the collection we had about fourteen of the very common local dormice, creatures that closely resemble European dormice, except that they are pale ash-grey, and have slightly more bushy tails. This colony of dormice lived in a cage together in perfect amenity, and in the evenings gave us a lot of pleasure with their acrobatic displays. There was one in particular that we could distinguish from all the others, for he had a tiny white star on his flank, like a minute cattle brand. He was a much better athlete than the others and his daring leaps and somersaults had earned our breathless admiration. Because of his circus-like abilities we had christened him Bertram.

One morning, as usual, I had let Cholmondely St. John out for his constitutional and he was behaving himself in an exemplary fashion. Then came a moment when I thought that Jacquie was watching him and she thought that I was. Cholmondely was always on the lookout for such opportunities. When we discovered our mistake and went in search of him we found we were too late. Cholmondely had amused himself by opening the doors of the dormouse sleeping compartments and then tipping the cage over so that the unfortunate rodents, all in a deep and peaceful sleep, cascaded onto the floor. As we arrived on the scene they were all rushing frantically for cover while Cholmondely, uttering small "Oooo's" of delight, was galloping around trying to stamp on them. By the time the ape had been caught and chastised there was not a dormouse in sight, for they had all gone to continue their interrupted slumbers behind our rows of cages. The entire collection had to be moved, cage by cage, so that we could recapture the dormice. The first one to break cover from behind a monkey cage was Bertram, who fled down the veranda hotly pursued by Bob. As he hurled himself at the flying rodent, I shouted a warning.

"Remember the tail. Don't catch it by the tail," I yelled.

But I was too late. Seeing Bertram wriggling his fat body behind another row of cages, Bob grabbed him by his tail, which was the only part of his anatomy easily grabbed. The result was disastrous. All small rodents, and

particularly these dormice, have very fine skin on the tail, so if you catch hold of it and the animal pulls away the skin breaks and peels off the bone of the tail like the finger of a glove. This is such a common thing among small rodents that I am inclined to think it may be a defence mechanism, like a lizard's dropping its tail when caught by an enemy. Bob knew this as well as I did, but in the excitement of the chase he forgot it, and so Bertram continued on his way behind the cage and Bob was left holding a fluffy tail dangling limply between finger and thumb.

Eventually we unearthed Bertram and examined him. He sat plumply in the palm of my hand, panting slightly, his tail, now pink and skinless, looking revoltingly reminiscent of an ox-tail before it enters a stew. As usual when this happens, the animal appeared to be completely unaffected by what would be, in the human, the equivalent of having all the skin suddenly ripped off one leg, leaving nothing but the bare bone and muscle. I knew from experience that eventually, deprived of skin, the tail would wither and dry, and then break off like a twig, leaving the animal none the worse off. In the case of Bertram, of course, the loss would be a little more serious, because he used his tail quite extensively as a balancing organ during his acrobatics, but he was so agile I did not think he would miss it much. But from our point of view Bertram was now useless, for he was a damaged specimen. The only thing to do was to amputate his tail for him and let him go. This I did, and then, very sorrowfully, we put him among the thick twining stems of the bougainvillæa that grew along the veranda rail. We hoped that he would set up house in the place and perhaps entertain future travellers with his acrobatic feats when he had grown used to having no tail.

He sat on a bougainvillæa stem, clutching it tightly

with his little pink paws, and looking about him through a quivering windscreen of whiskers. Then, very rapidly, and apparently with his sense of balance completely unimpaired, he jumped down on to the veranda rail and from there to the floor, then scurried across to the line of cages against the far wall. Thinking that perhaps he was a bit bewildered, I picked him up and returned him to the bougainvillæa. But as soon as I released him he did exactly the same thing again. Five times I put him in the bougainvillæa and five times he got down on to the veranda floor and made a bee-line for the cages. After that I got tired of his stupidity and carried him right down to the other end of the veranda, put him once more in the creeper, and left him, thinking that that would finish the matter.

On top of the dormouse cage we kept a bundle of cotton waste which we used to replenish their beds when they got too unhygienic, and that evening, when I went to feed them, I decided that they could do with a clean bed. So, after removing the extraordinary treasure trove that dormice like to keep in their bedrooms, I pulled out all the dirty cotton waste and prepared to replace it with clean. As I seized the bundle of waste on top of the cage, preparatory to ripping off a handful, I was suddenly bitten in the thumb. It gave me a considerable shock, for first of all I was not expecting it and secondly, for a moment, I thought it might be a snake. However, my mind was soon set at rest about the nature of my attacker, for as soon as I touched the cotton waste an indignant face was poked out of its depths and Bertram chittered and squeaked at me in what was obviously extremely indignant dormousese. Considerably annoyed with the wretched rodent, I hauled him out of his cosy bed, carried him along the veranda, and pushed him back into the bougainvillæa.

He clung indignantly to a stem, teetering to and fro and chittering furiously. But within two hours he was back in the bundle of cotton waste.

Giving up the unequal struggle, we left him there, but Bertram had not finished yet. Having beaten us into submission over the matter of accommodations, he then started to work on our sympathies in another direction. In the evening, when the other dormice came out of their bedroom and discovered their food plate with squeaks of surprise and delight, Bertram would come out of his bed and crawl down the wire front of the cage. There he would hang, peering wistfully through the wire, while the other dormice nibbled their food and carried away choice bits of banana and avocado to hide in their beds, a curious habit that dormice have, presumably to guard against night starvation. He looked so pathetic, hanging on the wire, watching the others stagger about with their succulent titbits, that eventually we gave in, and a small plate of food was placed on top of the cage for him. At last his cunning served its purpose: it seemed silly, since we had to feed him, to have him living outside, so we caught him and put him back in the cage with the others, where he settled down again as if he had never left, but, it seemed to us, he looked a trifle more smug than he used to do. But what other course could you adopt with an animal that refused to be let go?

Gradually we got everything under control. All the cages that needed repair were repaired, and each cage had a sacking curtain hung in front that could be lowered when travelling. The poisonous-snake boxes had a double layer of fine gauze tacked over them, to prevent accidents, and their lids were screwed down. Our weird variety of equipment, ranging from mincing machines to generators, from hypodermics to weighing machines, was packed away in crates and nailed up securely, and netting film tents were folded together with our giant tarpaulins. So everything was ready, and we awaited the fleet of trucks that was to take us down to the coast.

The night before they were due to arrive, the Fon came over for a farewell drink.

"Wah!" he exclaimed sadly, sipping his drink. "I sorry too much you leave Bafut, my friend."

"We get sorry too," I replied honestly. "We done have happy time here for Bafut. And we get plenty fine beef."

"Why you no go stay here?" inquired the Fon. "I go give you land for build one foine house, and den you go make dis your zoo here for Bafut. Den all dis European go come from Nigeria for see dis your beef."

"Thank you, my friend. Maybe some other time I go come back for Bafut and build one house here. Na good idea dis."

"Foine, foine," said the Fon, holding out his glass.

Down in the road below the rest house a group of the Fon's children were singing a plaintive Bafut song I had never heard before. Hastily I got out the recording machine, but just as I had it fixed up, the children stopped singing.

The Fon watched my preparations with interest. "You fit get Nigeria for dat machine?" he inquired.

"No, dis one for make record only, dis one no be radio."

"Ah!" said the Fon intelligently.

"If dis your children go come for up here and sing dat song I go show you how dis machine work," I said.

"Yes, yes, foine," said the Fon, and roared at one of his wives who was standing outside on the dark veranda. She scuttled down the stairs and presently reappeared, herding a small flock of shy, giggling children before her. I got them assembled round the microphone and then, with my fingers on the switch, looked at the Fon.

"If they sing now I go make record," I said.

The Fon rose majestically to his feet and towered over the group of children. "Sing," he commanded, waving his glass of whisky at them.

Overwhelmed with shyness, the children made several false starts, but gradually their confidence increased and they started to carol lustily. The Fon beat time with his whisky glass, swaying to and fro to the tune, occasionally bellowing out a few words of the song with the children. When the song came to an end, he beamed down at his progeny.

"Foine, foine, drink," he said, and as each child stood before him with cupped hands held up to his mouth, he poured a tot of almost neat whisky into the pink palms. While the Fon was doing this I wound back the tape and set up the machine for playback. Then I handed the earphones to the Fon, showed him how to adjust them, and switched on.

The expressions that chased themselves across the Fon's face were a treat to watch. First there was an expression of blank disbelief. He removed the headphones and looked at them suspiciously. Then he replaced them and listened with astonishment. Gradually, as the song progressed, a wide urchin grin of pure delight spread across his face.

"Wah! Wah! Wah!" he whispered in wonder. "Na wonderful, dis."

It was with the utmost reluctance that he relinquished the earphones so that his wives and councillors could listen as well. The room was full of exclamations of delight and the clicking of astonished fingers. The Fon insisted on singing three more songs, accompanied by his children, and then listening to the playback of each one, his delight undiminished by the repetition.

"Dis machine na wonderful," he said at last, sipping his drink and eyeing the recorder. "You fit buy dis kind of machine for Cameroons?"

"No, they no get um here. Sometime for Nigeria you go find um. Maybe for Lagos," I replied.

"Wah! Na wonderful," he repeated dreamily.

"When I go for my country I go make dis your song for proper record, and then I go send for you so you fit put um for dis your gramophone," I said.

"Foine, foine, my friend," he said. An hour later he left us, having embraced me fondly, and assuring us that he would see us in the morning before the trucks left. We were just preparing to go to bed, for we had a strenuous day ahead of us, when I heard the soft shuffle of feet on the veranda outside, and then the clapping of hands. I went to the door, and there on the veranda stood Foka, one of the Fon's elder sons, who bore a remarkable resemblance to his father.

"Hallo, Foka, welcome. Come in," I said.

He came into the room carrying a bundle under his arm, and smiled at me shyly. "De Fon send dis for you, sah," he said, and handed the bundle to me.

Mystified, I unravelled it. Inside was a carved bamboo walking stick, a small heavily embroidered skullcap, and a set of robes in yellow and black, with a beautifully embroidered collar.

"Dis na Fon's clothes," explained Foka. " 'E send um for you. De Fon 'e tell me say dat now you be second Fon for Bafut."

"Wah!" I exclaimed, genuinely touched. "Na fine ting dis your father done do for me."

Foka grinned delightedly at my obvious pleasure.

"Which side you father now? 'E done go for bed?" I asked.

"No, sah, 'e dere dere for dancing house."

I slipped the robes over my head, adjusted my sleeves, placed the ornate little skullcap on my head, grasped the walking stick in one hand and a bottle of whisky in the other, and turned to Foka.

"I look good?" I inquired.

"Fine, sah, na fine," he said, beaming.

"Good. Then take me to your father."

He led me across the great, empty compound and through the maze of huts towards the dancing house, where we could hear the thud of drums and the pipe of flutes. I entered the door and paused for a moment. The band in sheer astonishment stopped dead. There was a rustle of astonishment from the assembled company, and I could see the Fon seated at the far end of the room, his glass arrested halfway to his mouth. I knew what I had to do, for on many occasions I had watched the councillors approaching the Fon to pay homage or ask a favour. In dead silence I made my way down the length of the dance hall, my robes swishing round my ankles. I stopped in front of the Fon's chair, half crouched before him, and clapped my hands three times in greeting. There was a moment's silence and then pandemonium broke loose.

The wives and the council members screamed and hooted with delight. The Fon, his face split in a grin of pleasure, leaped from his chair and, seizing my elbows, pulled me to me feet and embraced me.

"My friend, my friend, welcome, welcome," he roared, shaking with gusts of laughter.

"You see," I said, spreading my arms so that the long sleeves of the robe hung down like flags, "you see, I be Bafut man now."

"Na true, na true, my friend. Dis clothes na my own one. I give for you so you be Bafut man," he crowed.

We sat down and the Fon grinned at me. "You like dis ma clothes?" he asked.

"Yes, na fine one. Dis na fine thing you do for me, my friend," I said.

"Good, good, now you be Fon same same for me." He laughed. Then his eyes fastened pensively on the bottle of whisky I had brought. "Good," he repeated. "Now we go drink and have happy time."

It was not until three-thirty that morning that I crawled tiredly out of my robes and crept under my mosquito net.

"Did you have a good time?" inquired Jacquie sleepily from her bed.

"Yes." I yawned. "But it's a jolly exhausting process being Deputy Fon of Bafut."

The next morning the trucks arrived an hour and a half before the time they had been asked to put in an appearance. This extraordinary circumstance—surely unparalleled in Cameroon history—allowed us plenty of time to load up. Loading up a collection of animals is quite an art. First of all you have to put all your equipment into the truck. Then the animal cages are put in towards the tailboard of the vehicle, where they will get the maximum amount of air. But you can't just push your cages in haphazardly. They have to be wedged in such a way that there are air spaces between each cage, and you have to make sure that the cages are not facing one another, or during the journey a monkey will push its hand through the wire of a cage opposite and get itself bitten by a civet; or an owl if placed opposite a cage of small birds will work them into such a state of hysteria (merely by being an owl and peering) that they will probably all be dead at the end of your journey. On top of all this you have to pack your cages in such a way that all the stuff that is liable to need attention en route is right at the back and easily accessible. By nine o'clock the last truck had been loaded and driven into the shade under the trees, and we could wipe the sweat from our faces and have a brief rest on the veranda. Here the Fon joined us presently.

"My friend," he said, watching me pour out the last enormous whisky we were to enjoy together, "I sorry too much you go. We done have happy time for Bafut, eh?"

"Very happy time, my friend."

"Shin-shin," said the Fon.

"Chirri-ho," I replied.

He walked down the long flight of steps with us, and at

the bottom shook hands. Then he put his hands on my shoulders and peered into my face. "I hope you an' all dis your animal walker good, my friend," he said, "and arrive quick-quick for your country." Jacquie and I clambered up into the hot, airless interior of the truck's cab and the engine roared to life. The Fon raised his large hand in salute, the truck jolted forward, and, trailing a cloud of red dust, we shuddered off along the road, over the golden green hills towards the distant coast.

The trip took three days, and was unpleasant and nerve-racking, as any trip with a collection of animals always is. Every few hours the trucks had to stop so that the small-bird cages could be unloaded and laid along the side of the road, and their occupants allowed to feed. Unless this was done the small birds would all die very quickly, for they seemed to lack the sense to feed while the truck was in motion. Then the delicate amphibians had to be taken out in their cloth bags and dipped in a local stream every hour or so, for as we got down into the forested lowlands the heat became intense, and unless this was done they would soon dry up and die. Most of the road surfaces were pitted with potholes and ruts, and as the trucks dipped and swayed and shuddered over them we sat uncomfortably in the front seats, wondering miserably what precious creature had been maimed or perhaps killed by the last bump. At one point we were overtaken by a heavy rainfall, and the road immediately turned into a sea of glutinous red mud that sprayed up from under the wheels like bloodstained porridge; then one of the trucks —an enormous four-wheel-drive Bedford—got into a skid from which the driver could not extricate himself, and ended on her side in the ditch. By means of an hour's digging round her wheels, and laying branches so that her tires could get a grip, we managed to get her out, and fortunately none of the animals was any the worse for the experience.

But we were filled with a sense of relief as the vehicles roared down through the banana groves to the port. Here the animals and equipment were unloaded and then stacked on the little flat-topped railway wagons used for conveying bananas to the side of the ship. These chugged and rattled their way through half a mile of mangrove swamp and then drew up on the wooden jetty where the ship was tied up. Once more the collection was unloaded and stacked in the slings, ready to be hoisted aboard. I went up onto the ship and made my way down to the forward hatch, where the animals were to be stacked, to supervise the unloading. As the first load of animals was touching down on the deck a sailor appeared, wiping his hands on a bundle of cotton waste. He peered over the rail at the line of railway trucks, piled high with cages; then he looked at me and grinned.

"All this lot yours, sir?" he inquired.

"Yes," I said, "and all that lot down on the quay."

He went forward and peered into one of the crates. "Blimey!" he said. "These all animals?"

"Yes, the whole lot."

"Blimey," he said again, in a bemused tone of voice, "you're the first chap I've ever met with a zoo in his luggage."

"Yes," I said happily, watching the next load of cages swing on board, "and it's my own zoo, too."

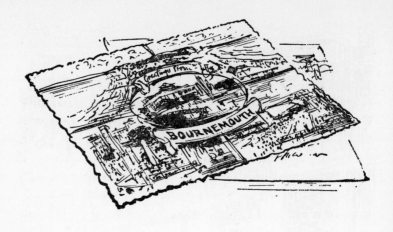

POSTCARD

Yes, bring the animals here. Don't know what the neigh-bours will say but never mind. Mother very anxious to see chimps so hope you are bringing them as well. See you all soon. Much love from us all.

Margo

8: A ZOO IN SUBURBIA

Most people who lived on our suburban road in Bournemouth could look out on their back gardens with pride, for each one resembled its neighbour's. There were minor differences, of course—some favoured pansies to sweet peas, or hyacinths to lupins—but basically they were all the same. Anyone looking out at my sister's back garden, however, would have been forced to admit that it was, to say the least, unconventional. In one corner was a huge marquee, from the interior of which came a curious chorus of squeaks, whistles, grunts, and growls. Alongside it stretched a line of Dexion cages from which glowered eagles, vultures, owls, and hawks. Next to them was a large cage containing Minnie, the chimp. On the remains of what had once been a lawn, fourteen monkeys rolled and played on long leashes, while in the garage frogs croaked, touracos called throatily, and squirrels gnawed loudly on hazelnut shells. At all hours of the day the fascinated, hor-

rified neighbours stood trembling behind their lace cur-
tains and watched as my sister, my mother, Sophie, Jacquie,
and I trotted to and fro through the shambles of the garden, carrying little pots of bread and milk, plates of
chopped fruit, or, what was worse, great hunks of gory
meat or dead rats. We had, the neighbours felt, taken un-
fair advantage of them. If it had been a matter of a crow-
ing cockerel, or a barking dog, or our cat having kittens in
one of their best flowerbeds, they would have been able to
cope with the situation. But the action of suddenly plant-
ing what amounted to a sizable zoo in their midst was so
unprecedented and unnerving that it took their breath
away, and so it was some time before they managed to rally
their forces and start to complain.

In the meantime I had started on my search for a zoo in
which to put my animals. The simplest thing to do, it oc-
curred to me, was to go to the local council and inform
them that I had the contents of a fine little zoo, and that
all I wanted them to do was let me rent or purchase a
suitable site on which to have it. Since I already had the
animals, it seemed to me in my innocence, they would be
delighted to help. It would cost them nothing, and they
would be getting what was, after all, another amenity for
the town. But the Powers That Be had other ideas.
Bournemouth is nothing if not conservative. There had
never been a zoo in the town ever since it had become a
town, and so they did not see why there should be one
now. This is what is known among local councils as progress. First they said that the animals would be dangerous;
then they said they would smell; and then, searching their
minds wildly for ideas, they said they did not have any
land anyway.

I began to get a trifle irritable. I am never at my best
when dealing with the pompous illogicalities of the offi-

cial mind. But I was beginning to grow worried in the face of such complete lack of cooperation. The animals were sitting in the back garden, eating their heads off and costing me a small fortune weekly in meat and fruit. The neighbours, now thoroughly indignant that we should not be conforming to pattern, kept bombarding the local health authorities with complaints, so that on an average of twice a week the poor inspector was forced to come up to the house, whether he wanted to or not. The fact that he could find absolutely nothing to substantiate the wild claims of the neighbours made no difference: if he received a complaint he had to come and investigate. We always gave the poor man a cup of tea, and he grew quite fond of some of the animals, even bringing his little daughter to see them. But what worried me most was that winter was nearly upon us, and the animals could not be expected to survive its rigours in an unheated marquee.

Then Jacquie had a brilliant idea. "Why not let's offer them to one of the big stores in town as a Christmas show?"

So I phoned up every big store in town. They were charming but unhelpful; they simply had not the space for such a show, however desirable. Then I phoned up the last on my list, the huge emporium owned by J. J. Allen. They, to my delight, evinced great interest and asked me to go and discuss it with them. And so "Durrell's Menagerie" came into being.

A large section of one of their basements was set aside, roomy cages were built with tastefully painted murals on the walls depicting a riot of tropical foliage, and the animals were moved in out of the cold and damp which had already started, into the luxury of brilliant electric light and a constant temperature. The charge for admission just covered the food bills, so the animals were warm, comfortable and well fed without being a drain on my resources. With this worry off my mind I could turn my attention once more to the problem of getting my zoo.

It would be wearisome to go into all the details of the frustration of this period, or to make a catalogue of the number of mayors, town councillors, park superintendents, and sanitary officers I met and argued with. Suffice it to say that I felt my brain creaking at times with the effort of trying to persuade supposedly intelligent people that a zoo in any town should be considered an attraction rather than anything else. The way everyone went on, one would have thought that I wanted to set off an atomic bomb on one of the piers.

In the meantime the animals, unaware that their fate hung in the balance, did their best to make life exciting for us. There was, for example, the day that Georgina the baboon decided that she wanted to see a little more of Bournemouth than the inside of J. J. Allen's basement. Fortunately it was a Sunday morning, so there was no one in the store; I dread to think what would have happened if there had been.

I was sipping a cup of tea, preparatory to going down to the store and cleaning and feeding the animals, when the telephone rang. Without a care in the world I answered it.

"Is that Mr. Durrell?" inquired a deep, lugubrious voice.

"Yes, speaking."

"This is the police 'ere, sir. One of them monkeys of yours 'as got out, and I thought I'd better let you know."

"Good God, which one is it?" I asked.

"I don't know sir, really. It's a big brown one. Only it looks rather fierce, sir, so I thought I'd let you know."

"Yes, thanks very much. Where is it?"

"Well, it's in one of the windows at the moment. But I

don't see as 'ow it'll stay there very long. Is it liable to
bite, sir?"

"Well, it may do. Don't go near it. I'll be right down,"
I said, slamming down the receiver.

The last thing I wanted was to get down there and find
a bloodstained constable. I grabbed a taxi and we roared
down to the centre of the town, ignoring all speed limits.
After all, I reflected, we were on police business of a sort.

As I paid off the taxi fare, the first thing that greeted
my eyes was the chaos in one of the big display windows of
Allen's. The window had been carefully set out to show
some articles of bedroom furniture. There was a large
bed, made up, a tall bedside light, and several eiderdowns
tastefully spread over the floor. At least, that was how it
had looked when the window dresser had finished it. Now
it looked as if a bomb had hit it. The light had been over-
turned and had burned a large hole in one of the eiderdowns; the bedclothes had been stripped off the bed, and
the pillow and sheets were covered with a tasteful pattern
of paw marks. On the bed itself sat Georgina, bouncing up
and down happily, and making ferocious faces at a crowd
of scandalized church-goers who had gathered on the pave-
ment outside the window. I went into the store and found
two enormous constables lying in ambush behind a barri-
cade of turkish towelling.

"Ah," said one with relief, "there you are sir. We didn't
like to try and catch it, see, because it didn't know us, and
we thought it might make it worse, like."

"I don't think anything could make that animal worse,"
I said bitterly. "Actually she's harmless, but she makes a
hell of a row and looks fierce. It's all bluff, really."

"Really?" said one of the constables, polite but uncon-
vinced.

"I'll try and get her in the window there if I can, but if

she breaks away I want you two to head her off. Don't, for the love of Allah, let her get into the china department."

"She came through the china department already," said one of the constables with gloomy satisfaction.

"Did she break anything?" I asked faintly.

"No, sir, luckily. She just galloped straight through. Me and Bill was chasing 'er of course, so she didn't stop."

"Well, don't let's let her get back in there. We may not be so lucky next time."

By this time Jacquie and my sister Margo had arrived in another taxi, so our ranks had now swelled to five. We should, I thought, be able to cope with Georgina between us.

I stationed the two constables, my sister, and my wife at suitable points guarding the entrance to the china department, and then went round and entered the window in which Georgina was still bouncing up and down on the ruined bed, making obscene faces at the crowd.

"Georgina," I said in a quiet but soothing voice, "come along then, come to Dad."

Georgina glanced over her shoulder in surprise. She studied my face as I moved towards her, and decided that my expression belied my honeyed accents. She gathered herself and leaped through the air, over the still smouldering eiderdown, and grabbed at the top of the great rampart of turkish towelling that formed the background of the window display. This, not having been constructed to take the weight of a large baboon hurtling through the air, immediately collapsed, and Georgina fell to the ground under a cascade of many-hued towelling. She struggled madly to free herself, and succeeded in doing so just as I flung myself forward to catch her. She gave a hysterical squawk and fled out of the window into the interior of the shop. I unravelled myself from the towelling and followed.

A piercing shriek from my sister told me of Georgina's whereabouts; my sister always tends to go off like a locomotive in moments of crisis. Georgina had slipped past my sister and was now perched on a counter, surveying us with glittering eyes, thoroughly enjoying the game. We approached her in a grim-faced body. At the end of the counter, suspended from the ceiling, hung a Christmas decoration made out of holly, tinsel, and cardboard stars. It was shaped somewhat like a chandelier, and seemed, as far as Georgina was concerned, the ideal thing to swing on. She poised herself on the end of the counter and as we ran forward she leaped up and grabbed at the decoration in a manner vaguely reminiscent of the elder Fairbanks. The decoration, not having been designed for this sort of treatment, promptly gave way, and Georgina fell to the ground, leaped to her feet, and galloped off, wearing a piece of tinsel over one ear.

For the next half hour we thundered through the deserted store, always with Georgina one jump ahead of us, as it were. She knocked down a huge pile of account books in the stationery department, paused to see if a pile of lace doilies was edible, and did a large and decorative puddle at the foot of the main staircase. Then, just as the constables were beginning to breath rather stertoriously, and I was beginning to despair of ever catching the wretched animal, Georgina made a miscalculation. Loping easily ahead of us, she came upon what looked like the perfect hiding place made out of rolls of linoleum arranged on end. She fled between the rolls and was lost, for the rolls had been arranged in the form of a hollow square, a three-sided trap from which there was no escape. Quickly we closed in and blocked the entrance to the linoleum trap. I advanced towards her, grim-faced, and she sat there and screamed wildly, begging for mercy. As I made a lunge to grab her she ducked under my hand, and as I swung round to prevent her escape I bumped into one of the massive rolls of linoleum. Before I could stop it this toppled forward like a gigantic truncheon and hit one of the constables accurately on the top of his helmet. As the poor man staggered backwards Georgina took one look at my face and decided that she was in need of police protection. She rushed to the still swaying constable and wrapped her arms tightly round his legs, looking over her shoulder at me and screaming. I jumped forward and grabbed her by her hairy legs and the scruff of her neck, and dragged her away from the constable's legs, still screaming piercingly.

"Cor!" said the constable, in a voice of deep emotion, "I thought I'd 'ad me chips that time."

"Oh, she wouldn't have bitten you," I explained, raising my voice above Georgina's harsh screams. "She wanted you to protect her from me."

"Cor!" he said again. "Well, I'm glad *that*'s over."

We put Georgina back in her cage, thanked the constables, cleared up the mess, cleaned and fed the animals, and then went home. But for the rest of that day, every time the phone rang I nearly jumped out of my skin.

Another animal that did its best to keep us on our toes was, of course, Cholmondely St. John, the chimp. To begin with, having established himself in the house and having got my mother and sister well under control, he then proceeded to get a nasty chill that rapidly developed into bronchitis. After recovering from this, he was still very wheezy, and so I decreed that he should, for the first winter at any rate, wear clothes to keep him warm. As he lived in the house with us he already was wearing plastic pants and paper nappies, so he was used to the idea of clothes. As soon as I had made this decision my mother, a delighted gleam in her eye, set to work, her knitting needles clicking ferociously, and in record time had provided the ape with a variety of woolly pants and jerseys, in the most complicated Fairisle patterns and of brilliant colourings. So Cholmondely St. John would loll on the window sill of the drawing room, nonchalantly eating an apple, clad in a different suit for each day of the week, completely ignoring the fascinated groups of local children that hung over our front gate and watched him absorbedly.

The attitude of people towards Cholmondely I found very interesting. Children, for example, did not expect him to be anything more than an animal with a curious resemblance to a human being, and with the ability to make them laugh. The adults who saw him, I'm afraid, were much less wise. On numerous occasions I was asked by apparently intelligent people whether he could talk. I always used to reply that chimps have, of course, a limited language of their own. But this is not what my questioners meant; they meant could he talk like a human being, could he discuss the political situation or the cold war or some equally fascinating topic.

But the most extraordinary question I was ever asked about Cholmondely was asked by a middle-aged woman on the local golf links. I used to take Cholmondely up there on fine days and let him scramble about in some pine trees, while I sat on the ground beneath, reading or writing. On this particular day Cholmondely had played for half an hour or so in the branches above me and then, growing bored, had come down to sit on my lap and see if he could inveigle me into tickling him. Just at that moment this strange woman strode out of the gorse bushes, and on seeing Cholmondely and me she stopped short. She displayed none of the surprise that most people evince at finding a chimpanzee in a Fairisle pullover occupying the golf links. She came closer and watched Cholmondely closely as he sat on my lap. Then she turned to me and fixed me with a gimlet eye.

"Do they have souls?" she inquired.

"I don't know, madam," I replied. "I can't speak with any certainty for myself on that subject, so you can hardly expect me to vouch for a chimpanzee."

"Um," she said, and walked off. Cholmondely had that sort of effect on people.

Having Cholmondely living in the house with us was, of course, a fascinating experience. His personality and intelligence made him one of the most interesting animals I have ever kept. One of the things about him that impressed me most was his memory, which I thought was quite phenomenal.

I possessed at that time a Lambretta and sidecar, and I decided that, providing Cholmondely sat well in the sidecar and didn't try to jump out, I would be able to take him for excursions into the countryside. The first time I introduced him to it, I took him for a round trip of the golf links, just to see how he would behave. He sat there with the utmost decorum, watching the passing scenery with a regal air; apart from a tendency to lean out of the sidecar and try to grab any cyclist we overtook, his behaviour was exemplary. Then I drove the Lambretta down to the local garage to have her filled up with petrol. Cholmondely was as fascinated with the garage as the garage man was with Cholmondely. The ape leaned out of the sidecar and watched the unscrewing of the petrol tank absorbedly, and the introduction of the hose and splash and gurgle of the petrol made him "Ooo" softly to himself in astonishment. Now, a Lambretta can travel an incredible distance on a very small amount of petrol and, as I did not use it a great deal, it was about two weeks before she needed filling up again. We had just come back from a local water mill where we had been visiting Cholmondely's friend, the miller. This kind man, a great admirer of Cholmondely's, always had a brew of tea ready for us, and we would sit in a row above the weir, watching the moorhens paddling by, sipping our tea, and meditating. On the way home from this tea party I noticed that the Lambretta was getting low on fuel, so we drove down to the garage.

As I was passing the time of day with the garage man, I noticed that he was gazing over my shoulder, a somewhat stupefied expression on his face. I turned round quickly to see what mischief the ape was up to. I found that Cholmondely had climbed out of the sidecar onto the saddle and was very busy trying to unscrew the cap of the petrol tank, preparatory to having us filled up. Now, I considered this to be quite a feat of memory. First, he had seen the filling-up process only once, and that had been two weeks previously. Second, he had remembered, out of all the various gadgets on the Lambretta, which was the correct one to open in these circumstances. I was almost as impressed as the garage man.

But Cholmondely impressed me most, not only with his memory but with his powers of observation, on the occasions when I had to take him up to London, once to appear on television and later for a lecture. My sister drove me up to London, while Cholmondely sat on my lap and watched the passing scenery with interest. About halfway to our destination I suggested that we stop for a drink. One had to be rather careful about the pubs you stopped at when you had Cholmondely with you, for it was not every landlord that appreciated a chimpanzee in his private bar. Eventually we found a pub that had a homely look about it, and stopped there. To our relief, and Cholmondely's delight, we found that the woman who ran the

pub was a great animal lover, and she and Cholmondely took an immediate fancy to each other. He was allowed to play catch as catch can among the tables in the bar, he was stuffed with orange juice and potato chips, he was even allowed to get up on the bar itself and do a war dance, thumping his feet and shouting, "Hoo . . . hoo . . . hoo." In fact he and the landlady got on so well that he was very reluctant to leave the place at all. If he had been a Royal Automobile Club inspector he would have given the pub twelve stars.

Three months later I had to take Cholmondely up for the lecture; by this time I had forgotten all about the pub in which he had had such a good time, for we had, since then, been in many other licensed establishments which had given him a warm welcome. As we drove along, Cholmondely, who was sitting on my lap as usual, started to bounce up and down excitedly. I thought at first he had seen a herd of cows or a horse—animals in which he had the deepest interest—but there was not a farm animal to be seen. Cholmondely went on bouncing, faster and faster, and presently started *Oo*ing to himself. I still could not see what it was that was exciting him. Then his *Oo*ing rose to a screaming crescendo, and he leaped about on my lap in an ecstacy of excitement; we rounded a corner and there, a hundred yards ahead, was his favourite pub. Now this meant that he had recognized the countryside we were passing through, and had connected it with his memory of the good time he had had in the pub, a mental process which I had not come across in any other animal. Both my sister and I were so amazed by this that we were very glad to stop for a drink, and let Cholmondely renew his acquaintance with his friend the landlady, who was delighted to see him again.

In the meantime I was still continuing my struggle to

start my zoo, but my chance of success seemed to recede farther and farther each passing day. The collection had to be moved from J. J. Allen's, of course, but here Paignton Zoo came to my rescue. With extreme kindness the authorities allowed me to put my collection up with them, on deposit, until such time as I could find a place of my own. But this, as I say, began to seem more and more unlikely. It was the old story. In the initial stages of a project, when you need people's help most, it is never forthcoming. The only thing to do, if you can, is to go ahead and accomplish it by yourself. Then, when you have made a success of it all, the people who would not help you get it launched gather round, slap you on the back, and offer their assistance.

"There must be an intelligent local council *somewhere*," said Jacquie one evening, as we pored over a map of the British Isles.

"I doubt it," I said gloomily, "and anyway I doubt whether I have the mental strength to cope with another round of mayors and town clerks. No, we'll just have to get a place and do it ourselves."

"But you'll have to get their sanction, and then there's Town and Country Planning and all that."

I shuddered. "What we should really do is go to some remote island in the West Indies, or somewhere," I said, "where they're sensible enough not to clutter their lives with all this incredible red tape."

Jacquie moved Cholmondely St. John from the portion of the map he was squatting on. "What about the Channel Isles?" she asked suddenly.

"What about them?"

"Well, they're a very popular holiday resort, and they've got a wonderful climate."

"Yes, it would be an excellent place, but we don't know

anyone there," I objected, "and you need someone on the
spot to give you advice in this sort of thing."

"Yes," said Jacquie, reluctantly, "I suppose you're
right."

So, reluctantly (for the idea of starting my zoo on an
island had a very strong appeal for me) we forgot about
the Channel Islands. It was not until a few weeks later
when I happened to be in London and was discussing
my zoo project with Rupert Hart-Davis that a gleam of
daylight started to appear. I confessed to Rupert that
my chances of having my own zoo now seemed so slight
that I was on the verge of giving up the idea altogether.
I said that we had thought of the Channel Islands, but that
we had no contact there to help us. Rupert sat up and,
with an air of a conjurer performing a minor miracle, said
he had a perfectly good contact in the Channel Islands (if
only he was asked) —a man, moreover, who had spent his
whole life in the islands and would be only too willing
to help us in any way. His name was Major Fraser, and
that evening I phoned him. He did not seem to find it
a bit unusual that a complete stranger should phone him
up and ask his advice about starting a zoo, which made me
warm to him from the start. He suggested that Jacquie
and I should fly across to Jersey and he would show us
round the island, and give us any information he could.

So we flew to Jersey. As the plane came in to land, the
island seemed like a toy continent, a patchwork of tiny
fields, set in a vivid blue sea. A pleasantly carunculated
rocky coastline was broken here and there with smooth
stretches of beach, along which the sea creamed in ribbons.
As we stepped out onto the tarmac the air seemed warmer,
and the sun a little more brilliant. I felt my spirits rising.

In the car park Hugh Fraser awaited us. He was a tall,
slim man, wearing a narrow-brimmed trilby tilted so far

forward that the brim almost rested on his aquiline nose. His blue eyes twinkled humorously as he shepherded us into his car and drove us away from the airport. We drove through St. Helier, the capital of the island, which reminded me of a sizable English market town, so it was something of a surprise to find, at a crossroads, a policeman in a white coat and white helmet directing the traffic. It suddenly gave the place a faintly tropical atmosphere. We drove through the town and then out along narrow roads with steep banks, where the trees leaned over and entwined branches, making a green tunnel. The landscape, with its red earth and rich green grass, reminded me very much of Devon, but the landscape was a miniature one, with tiny fields, narrow valleys stuffed with trees, and small farmhouses built of the beautiful Jersey granite, which contains a million autumn tints in its surface where the sun touches it. Then we turned off the road, drove down a long drive, and suddenly before us was Hugh's home, Les Augres Manor.

The Manor was built like an E without the centre bar; the main building was in the upright of the E, while the two cross pieces were wings of the house, ending in two massive stone arches which allowed access to the courtyard. These beautiful arches were built around 1660 and, like the rest of the building, were of the lovely local granite. Hugh showed us round his home with obvious pride; we saw the old granite cider press and cow sheds, the huge walled garden, the small lake with its tattered fringe of bullrushes, the sunken water meadows with the tiny streams trickling through them. At last we walked slowly back under the beautiful archways and into the courtyard, flooded with sunshine.

"You know, Hugh, you've got a wonderful place here," I said.

"Yes, it is lovely . . . I think one of the loveliest manors on the island," said Hugh.

I turned to Jacquie. "Wouldn't it make a wonderful place for our zoo?"

"Yes, it would," agreed Jacquie.

Hugh eyed me for a moment. "Are you serious?" he said.

"Well, I *was* joking, but it would make a wonderful site for a zoo. Why?" I asked.

"Well," said Hugh, thoughtfully, "I'm finding the upkeep too much for my resources, and I want to move to the mainland. Would you be interested in renting the place?"

"Would I?" I said. "Just give me the chance."

"Come inside, dear boy, and we'll discuss it," said Hugh, leading the way across the courtyard.

So, after a frustrating year of struggling with councils and other local authorities, I had gone to Jersey, and within an hour of landing at the airport I had found my zoo.

THE LAST WORD

My zoo in Jersey has now been open to the public for nearly a year. We are probably the newest zoo in Europe and, I like to think, one of the nicest. We are small, of course (at the moment we have only about six hundred and fifty mammals, birds, and reptiles), but we will continue to expand. Already we have on show a number of creatures which no other zoo possesses, and we hope, in the future, when funds permit, to concentrate on those species which are threatened with extinction.

Many of the animals on show are ones I collected myself. This is, as I said before, the best part of having one's own zoo, to be able to bring the animals back for it, to watch their progress, to watch them breed, to be able to go out and visit them at any hour of the day or night. But also I hope that, in a small way, I am interesting people in animal life and in its conservation. If I accomplish this I will consider that I have achieved something worth while. And if I can, later on, help even slightly towards preventing an animal from becoming extinct, I will be content.

ACKNOWLEDGMENTS

I would like to express the gratitude of all members of the expedition party for the overwhelming support that we received from British industry in the way of equipment, without which the trip would have been far more difficult and the results not so satisfying. This was the first time that I had ever asked for help, and we were absolutely overwhelmed by the response and unstinting generosity of the various firms concerned.

We also acknowledge our debt of gratitude to the following individuals:

LONDON: Mr. Miles, of Grindlay's Bank Shipping Department, without whose efforts no members of the expedition would ever have arrived in the Cameroons.

CAMEROONS

Victoria: Mr. Eric Saward, Acting Manager, U.A.C., and his wife, Sheila, who generously welcomed us to the Cameroons.

Mr. Mac Carney, Manager, U.A.C., who went out of his way to help us.

Mr. Walker, of Elders and Fyffes Ltd., who saw that all food for the animals was safely put upon the ship.

Mr. Dudding, Assistant Commissioner, for all his help in arranging all our permits to catch animals.

Mr. Austin, of the Agricultural Co-operative, who most kindly sent a large truck all the way up from the coast to Bafut to ensure that both we and our animals caught the ship on time.

Kumba: Dr. William Crewe, who so lavishly entertained both us and our animal cargo on our way down to the coast.

Mr. Gordon, Manager, U.A.C., who supplied us with a four-wheel drive Bedford truck to take our animals down to the coast.

Mamfe: Mr. John Henderson, Manager, U.A.C., for whom our gratitude knows no bounds.

Mr. John Topham, who invited both us and our animals to invade his house at the dead of night and did everything he could to assist us. He also provided a truck to take the animals down to the coast.

Mr. John Thrupp, District Officer of the Mamfe Division, who bore our complaints and protests with fortitude.

Mr. Martin Davis, Forestry Officer, who helped us in every way and brought us Tavy, our second black-footed mongoose.

Bamenda: Dr. Paul Gebauer, of the Cameroons Baptist Mission, who, as on previous expeditions, suffered much at our hands yet always welcomed us.

Mr. Brandt, Manager, U.A.C., and his wife Rona, who did everything they could to make our stay in Bamenda enjoyable.

Mr. Shadock, A.D.O., who helped in many ways to smooth our departure.

Mr. Macfarlane, Veterinary Officer of the Cameroons, who gave us invaluable assistance with our animal charges.

Mr. Stan Marriot, of the Agricutural Department, who recharged our camera batteries and repaired our Land-Rover on countless occasions.

Mr. Dennison, Manager, U.A.C., who helped us in any way he could.

Tiko: Mr. Bowerman, of C.D.C., who made all arrangements for us to stay in the rest house prior to our sailing.

Our thanks also to the captain, officers, and men of the M.V. *Nicoya,* and in particular to Mr. Terrance Huxtable, the chief steward, who bore with us and our animals with great fortitude and understanding.

Last of all we would like to thank our good friend the Fon of Bafut, for giving us "a happy time."